Electric Primary Batteries
A Practical Guide To Their Construction and Use

by Bernard E. Jones

with an introduction by Roger Chambers

Self Reliance Books

Get more historic titles on animal and stock breeding, gardening and old fashioned skills by visiting us at:

http://selfreliancebooks.blogspot.com/

Introduction

I am pleased to present yet another title on Historical Science.

The work is in the Public Domain and is re-printed here in accordance with Federal Laws.

As with all reprinted books of this age that are intended to perfectly reproduce the original edition, considerable pains and effort had to be undertaken to correct fading and sometimes outright damage to existing proofs of this title. At times, this task is quite monumental, requiring an almost total "rebuilding" of some pages from digital proofs of multiple copies. Despite this, imperfections still sometimes exist in the final proof and may detract from the visual appearance of the text.

I hope you enjoy reading this book as much as I enjoyed making it available to readers again.

Roger Chambers

PREFACE

THE present work is believed to meet the need for a simple, practical book on primary battery construction and use, and is addressed to all electrical workmen and students, both amateur and professional. It is based upon contributions to " Work " by electrical authorities, including A. H. Avery, A.I.E.F., I. W. Chubb, A.I.E.E., " Henwick," " Telephonist," the late George E. Bonney, and others ; and in addition it includes much matter from my own pen. Should further information on any particular point be required, it will be readily furnished through the " Questions and Answers " columns of " Work."

B. E. J.

CONTENTS

LIST OF ILLUSTRATIONS

ELECTRIC PRIMARY BATTERIES

CHAPTER I

ELEMENTARY PRINCIPLES, CHIEF TERMS, ETC.

THE electric battery works on what is known as the Galvanic or Voltaic principle. Galvani, in 1790, made the fundamental experiment which led to Volta, in 1793, undertaking further investigations, and ultimately inventing the primary battery. His first battery, the " Crown of Cups " (Fig. 1) consisted of a number of glass vessels containing salt water. The bent metallic strips end in pieces of copper and

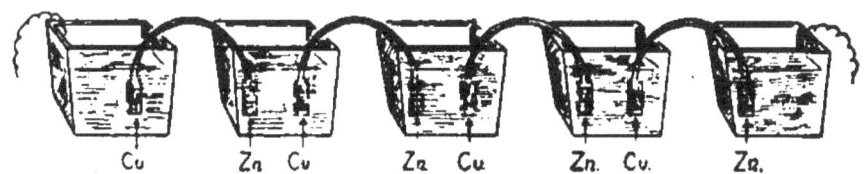

Fig. 1.—Volta's " Crown of Cups."

zinc respectively, the copper in one vessel and the zinc in the next. On running a wire from each of the terminal metals in the outer vessels to a muscle in a freshly killed frog, muscular contractions were produced, these contractions becoming more violent on increasing the number of the cups or cells.

One of these cells constitutes the simplest practical primary battery. Have a plate of copper and one of zinc standing in a vessel of acidulated water, as in Fig. 2. A wire is made to connect the two plates and so provide a path for the external circuit, the current being said to flow from the copper plate to the zinc in the external circuit, and from the zinc to the copper in the internal circuit.

Volta, in 1800, made his battery more compact by adopting the arrangement shown in Fig. 3; this is the " pile," from which the French term for any primary battery is derived. At the bottom of the pile is a disc of copper, on which is placed a disc of zinc, then a piece of cardboard moistened with acidulated water; this sequence of copper, zinc, and card is repeated as many times as convenient. A strip of copper connects the bottom copper plate to a vessel of acidulated water, and a second strip, or wire, connects the top zinc to a similar vessel. The

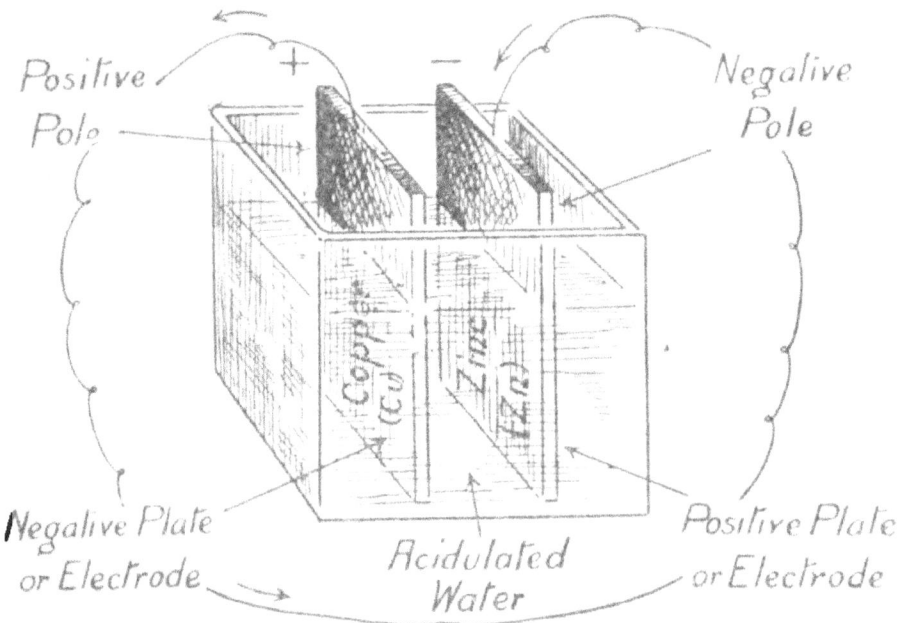

Fig. 2.—The Simplest Practical Cell.

connections to the apparatus, etc., are made from these two vessels. It is a convenience to replace this clumsy system of connections with terminal screws soldered on to the top and bottom discs.

The terms " cell " and " battery " are used indiscriminately. Obviously the term " battery " implies two or more cells connected together, but it frequently means nothing more than a single cell. The confusion is not of much consequence.

The direction of the current is of great importance. In theory, as already stated, the current always flows from the copper to the zinc outside the cell, and from zinc to copper inside, thus completing the circuit (*see* Fig. 2). This refers to the passage of a positive current, because it can be proved that the dry upper end of the copper plate is positively electrified and at a higher potential than the dry upper end of the zinc which is negatively electrified. There is always a difference of potential between a metal and the acidulated water in which it is

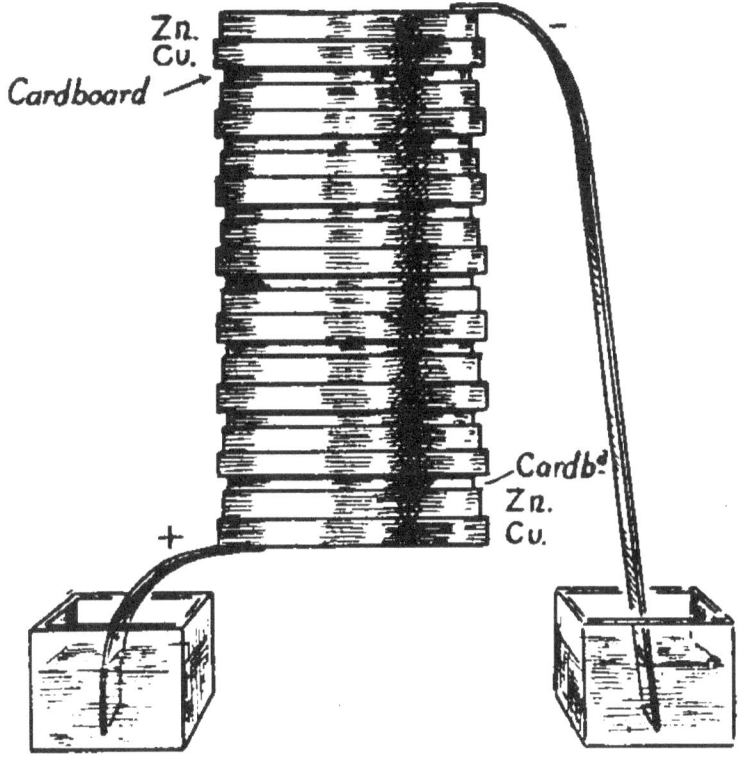

Fig. 3.—Volta's Pile.

immersed; the two are electrified, the water positively and the metal negatively. The amount of the difference of potential between acid and metal varies with the nature of the latter. Thus, curiously, the dry upper parts of both plates in a simple cell, when the connecting wire is removed, are negatively electrified, but owing to the difference between the metals they are not electrified equally, and it follows that on connecting them externally there is a difference of potential between them, and a positive current flows from the more highly charged of the two plates. The values change with each change of metal or of liquid, but there is always voltaic action while two different conductors are immersed in a liquid that can act chemically upon at least one of them.

The terms "positive" and "negative" may lead to confusion, unless the following is thoroughly grasped. There is assumed to be a flow of positive electricity from the exposed part of the copper to the zinc; thus, the terminal to the copper plate forms the positive *pole*, and that to the zinc plate the negative *pole* (*see* Fig. 2). But inside the battery, in completion of the circuit, the current is assumed to flow from zinc to copper; thus the immersed zinc plate is the positive *electrode*, and the immersed copper plate the negative *electrode* (*see* Fig. 2).

Current is assumed to flow both *from* the battery and *to* the battery; it leaves at the copper plate and re-enters at the zinc. The liquid, known as the electrolyte, is decomposed at the places where the current enters and leaves. The following action takes place when the external circuit is complete :—

(1) The current entering at the zinc plate decomposes the water in the electrolyte, forming oxygen; this combines with the zinc plate, forming zinc oxide which, in the presence of the sulphuric acid in the electrolyte, becomes zinc sulphate.

(2) The hydrogen is carried to the copper plate but cannot, under these conditions, combine with it, so it simply adheres, the plate ultimately becoming entirely coated with the hydrogen.

Polarisation.—After the cell has been working for some time, the "couple," as the two different metallic plates are known, is no longer what it was at starting. As a result of the above actions, the zinc and copper have become changed, more or less, to zinc and hydrogen, with the result that the difference of potential (otherwise, the pressure of the current) drops by about 25 per cent. In addition, the presence of the hydrogen increases the resistance to the passage of the current, and the yield of energy is still further reduced. This deterioration of the cell is known as " polarisation," and in the many attempts made to eliminate or, at least, minimise it have been produced most of the great variety of primary batteries known to-day.

Local Action.—Theoretically, when the external circuit is broken, all action ceases in a primary battery; but in practice some amount of local action is found to exist. Particles of iron or other metals, present as impurities in the zinc electrode, act themselves as miniature electrodes giving rise to electric currents which cause the zinc to be dissolved away in the neighbourhood of the impurities. The defect could be overcome by using chemically pure zinc, but this would be too expensive for ordinary use. The most practicable method is to amalgamate the zincs, to do which they are cleaned in sulphuric acid and then placed in mercury or rubbed with it. The mercury amalgamates with the surface of the zinc, which acts electrically equally as well as chemically pure zinc.

Local action is caused in certain cells by copper crystals which, after the cells have been in use for some time, separate out, pass through the porous pot, and come into direct contact with the zinc. The obvious remedy is thorough cleaning and re-making up of the cells.

The Electric Current.—Attention will now be directed to certain elementary matters which must be mastered by the student.

Somewhat in the sense that water flows through a pipe, electricity is assumed to flow as a current through a suitable metallic conductor.

An *electrical conductor* is any substance that offers but a slight resistance to the passage of an electric current. Theoretically, there is no perfect conductor—that is, there is no substance that is entirely without resistance. In general, good conductors of heat are also good conductors of electricity. The conductivity of a substance is its capability of conducting an electric current. Conductivity is said to be the reciprocal of resistance; in explanation of this, if one substance has a resistance of 4 ohms, and another substance has a resistance of 8 ohms, their respective conductivities are $\frac{1}{4}$ and $\frac{1}{8}$; in other words, the conductivity of the former substance is twice that of the latter substance. Conductors in the order of their decreasing conductivities are silver, copper, gold, aluminium, zinc, platinum, iron, tin, lead, German silver, mercury, charcoal, acids, and water. Copper and aluminium are commonly used for the purpose. On the other hand, an *insulator* is any substance that offers much resistance to the passage of an electric current. Theoretically, there is no substance that is a perfect insulator—that is, there is none that offers so great a resistance as to obstruct entirely the passage of a current of electricity. In general, poor conductors of heat are insulators. A list of insulators in the order of decreasing conductivities is as follows : Oils, porcelain, wool, silk, resin, gutta-percha, shellac, ebonite, paraffin, glass, and dry air. Porcelain, gutta-percha, ebonite and glass are among the commonest insulators used.

The quantity of a current that flows through a conductor is measured in amperes; the pressure, difference of potential, or electro-motive force (E.M.F.) is measured in volts; and the resistance which it has to overcome is measured in ohms. Thus, the ampere is the (practical) unit of quantity; the volt of pressure or E.M.F. ; and the ohm, of resistance. By a simple rule known as Ohm's law, any one of these three can be determined when the two others are known. The law may be popularly expressed in the following way :—

Amperes = volts divided by ohms.
(Current) = (E.M.F.) (resistance).

It follows from this, that

Volts = amperes multiplied by ohms.
Ohms = volts divided by amperes.

There is an easy method of committing this important rule to memory. Let the letter E represent volts (E.M.F.); C, amperes (current); and R, ohms (resistance). Then memorise the following simple expression :—

$$\frac{E}{C \times R}$$

(E divided by C R multiplied together). When any two of the above factors are known, the third is obtained by simple division or multiplication; blot out with the finger the required factor, and the remainder of the expression indicates the quantity.

To find E, multiply C by R.
To find C, divide E by R.
To find R, divide E by C.

The watt, the unit of power, is found by multiplying volts by amperes; 1,000 watts = 1 kilowatt.

CHAPTER II

CARBON-ZINC CELLS—THE BICHROMATE OR CHROMIC ACID

The Single-fluid Bichromate Battery

THIS is made in a variety of forms, all that is essential being a rod of carbon and another of zinc in an acidulated solution of bichromate of potash. However, the most usual commercial form is Grenet's flask or bottle battery (*see* Fig. 4), which has two carbon plates coupled together forming one element, with a plate of zinc for the other element sandwiched between them, separated by a small gap from the carbons, the whole being immersed in the bichromate of potash or chromic acid exciting solution. In the better class of bottle batteries the zinc is made to slide up out of the solution when not in use, as this preserves it much longer. The carbons are not acted upon by the solution, but the zinc is gradually consumed, and requires occasional renewal. The E.M.F. is approximately two volts, and the current is fairly constant; it can be made quite so by gradually lowering the zinc more and more into the solution as the current shows signs of falling off, as this reduces the internal resistance.

The arrangement of a zinc plate suspended between two carbon plates with only a slight space between them effects a reduced internal resistance, and also utilises both faces of the zinc plate. As a consequence, such a battery is capable of yielding a more powerful current than one in which there is only one carbon plate to each zinc.

When replenishing a bichromate cell the first thing to do is to empty out all the old solution and thoroughly wash out the battery with clean water. Cleanliness is essential with all types of batteries to get the best results, and more particularly with this one. If the zincs are much eaten away they should be renewed at the same time, and then a fresh exciting solution supplied. The most thorough method is to soak the carbon plates in hot water. rub them with a hard brush, and allow to drain,

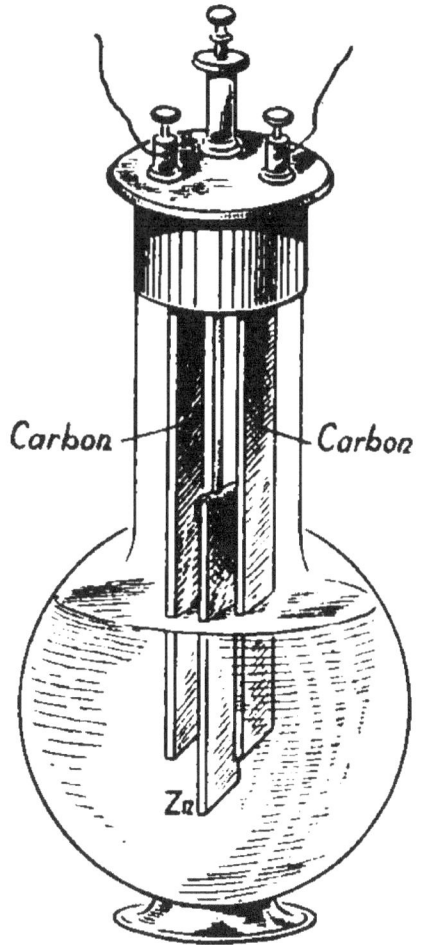

Fig. 4.—Grenet's Flask or Bottle Bichromate Cell.

If the zincs are not clean and bright, they should be re-amalgamated. To do this, have ready in a shallow dish a dilute solution of sulphuric acid, to which is added some mercury. First dip the zinc plates in dilute sulphuric acid for a few minutes, and then transfer them to the shallow dish. Then

B

with a clean old paint brush, or some tow fastened on a stick, rub the mercury over the zinc plates until they are clean and bright.

To make the solution to be used in all forms of single-fluid bichromate batteries, dissolve 3 oz. of bichromate of potash in 1 pt. (20 oz.) of hot rainwater, and set it aside until quite cold. Then add to it slowly, in a fine stream, or drop by drop, whilst being stirred, 3 oz. of strong sulphuric acid. Should the solution become warm by adding the acid, it must be set aside until quite cold before it is put in the cell. If this is not done, the zinc plate will be violently attacked, even when well coated with mercury. The zinc element must be lifted out clear of the solution when the battery is not in use, to keep this and the zinc in working order. As the battery is worked, the colour of the solution will alter from orange to greenish brown, then to a blackish-brown tint, denoting exhaustion, when it must be renewed, and the zinc freshly coated with mercury.

Another suitable solution may be made up as follows :—Dissolve 2 oz. of bichromate of potassium in 20 oz. of hot rainwater ; add slowly 5 oz. of strong sulphuric acid, stirring constantly, and dissolve in this mixture, while it is hot, $\frac{1}{2}$ oz. of bisulphate of mercury. Allow the solution to cool before using. When the solution in the cell turns green, it should be renewed.

The vessel in which the solution is mixed should be of stoneware, because the heat generated would crack glass.

The above solutions produce chromic acid in the course of their chemical activity ; but that acid may be supplied in the first case, if desired, in the place of the bichromate. Chromic acid is preferred by some as being a better depolariser, enabling the battery to give a steadier current. The proportions in this case are :—Water, 20 parts ; chromic acid,

3 parts; sulphuric acid, 1½ parts; and chlorate of potash, 1½ parts, all by weight. Other suitable proportions are :—Water, 20 parts; chromic acid, 3 parts; sulphuric acid, 3 parts; and chlorate of potash, ¼ part. The chromic acid and chlorate of potash must be dissolved in the water first, then the sulphuric acid in a fine stream or drop by drop, the solution being allowed to get quite cold before putting it in the cells. Fully eight hours' good work should be got out of the battery before it needs renewing.

There is no doubt that chromic acid batteries are among the most powerful of the primary batteries. If properly made and worked they furnish strong currents for fairly long periods, and are invaluable for experimental work. But the cells must be of glass, the zinc elements must be kept well coated with mercury, all the elements must be lifted out of the solution when not in use, and the solutions must be properly prepared.

The principal defects of single-fluid bichromate cell are :—(1) Rapid fall of current; (2) local action; (3) evaporation and consequent crystallisation. The first defect is minimised by the use of the porous pot and of a second fluid (as described later); the second, by thorough and constant amalgamation of the zincs, and provision for raising the electrodes out of the electrolyte when not in use; and the third, by using a closed or covered cell. It is not difficult to provide for the covering in of a home-made cell. Assume that the outer jar of an old Leclanché cell is used (see Fig. 12, p. 36). Carve from a piece of hard wood a cover to fit the inside of the mouth of the glass cell, with a flange to rest on the rim. Soak this in melted paraffin until the wood is saturated. The zinc and carbon elements are to be suspended from the under side of this cover. The carbon strips have their top ends coppered and soldered to tangs, which come up through the cover to binding screws above. The zinc strip, suspended between the carbons, is soldered to a brass rod, which slides in a brass tube connected to another binding screw. By this means the zinc is drawn up out of the solution when not in use.

Making a Bottle Battery.—Practical instruction on the making of a bottle bichromate cell will now be given. One of the most handy forms of this battery is a single cell made with a bottle so shaped as to be easily handled by the operator. This is the type already illustrated in Fig. 4. The bottle used must be large enough to hold ½ pint or more of water, and the mouth must be wide enough to. take the strips of carbon and zinc. Special glass bottles, similar to that shown, are made and sold for the purpose. An ordinary pickle or jam bottle, or the glass cell of a Leclanché battery may be utilised if desired. Unless the space for the battery is limited, a larger size is not objectionable, as the charge will then last longer, and the current will be stronger than from a small cell. The plates must be chosen to fit the cell. Two carbon plates wide enough to fit the neck of the bottle, and long enough to reach from the cover to within ½ in. of the cell bottom, will be required, and one zinc plate of the same width but only half the length of the carbon plates. The zinc plate will be suspended in the cell between the two carbon plates, as shown in Fig. 4, with not less than ¼ in. clear space on each side. If the zinc touches the carbons, the battery is useless. The plates may be ¼ in. thick ; the zinc plate should certainly not be thinner.

The top of each carbon plate, to the depth of ½ in., must be coated with copper, to form a connection with the solder employed to join the plate to a strip of brass or copper. The coppering process may be done as follows :—Make a pint of saturated solution of copper sulphate in rainwater, and put it in a stoneware jar. Stand in this a small porous cell nearly filled with dilute sulphuric acid. In this have a rod of zinc coated with mercury. Twist a piece of No. 22 copper wire around one end of the carbon plate, and attach the other end of the wire to the zinc rod; then immerse the other end of the carbon plate in the copper solution to the depth of ½ in., supporting the wire and carbon on a strip of wood. In a few minutes the carbon will receive a coat of copper, and at the end of an hour it will be thick enough for the purpose in hand. Well rinse it in hot water, and dry by rubbing in hot sand or sawdust. A small strip of copper or brass, bent to the form of a clip, may now be soldered to the coppered end of the carbon, after which the coppered part and connection should be coated with paraffin, varnish, or brunswick black. The two carbon plates, thus prepared, may now be set aside until the zinc plate has been got ready.

The upper edge of the zinc plate must be soldered to a connecting shank of brass or copper, and this shank must fit loose in a connecting socket fixed on the cover of the cell. This shank will be used to slide the zinc plate up and down in the solution as required. It must therefore be long enough to stand at least ¾ in. above the cover when the zinc is touching the bottom of the cell. A square rod will be the best shape, and the lower end should form a clip into which the zinc is fitted and soldered. This joint should then be coated with brunswick black, or with an insulating varnish not easily attacked by acid. The zinc must next be coated with mercury. This is best done in a stoneware pie-dish containing the mercury and a strong solution of sulphuric acid, the zinc being rubbed with a wisp of tow whilst wiping the mercury over it. The zinc must be thus coated every time the battery is cleaned and recharged. This rule applies to the zinc element of all bichromate batteries.

The cover of the battery is made of hard wood

in the form of a bung with a projecting rounded flange about ¼ in. thick, to make a kind of platform for the fittings. Two holes must be made in this

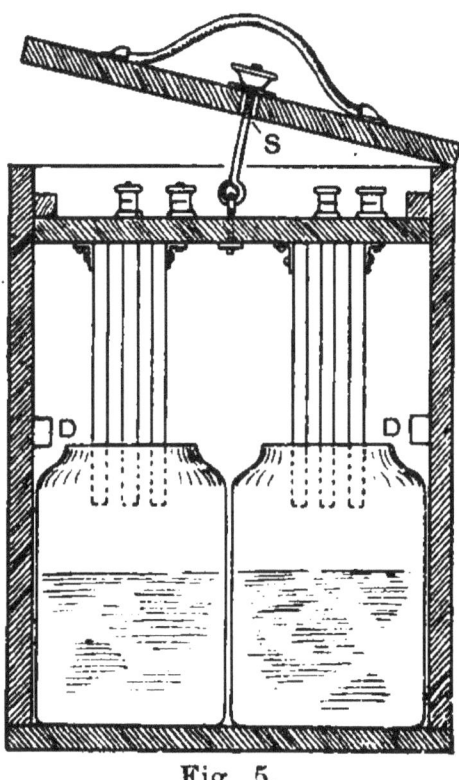

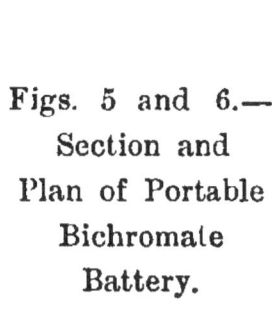

Figs. 5 and 6.— Section and Plan of Portable Bichromate Battery.

Fig. 5.

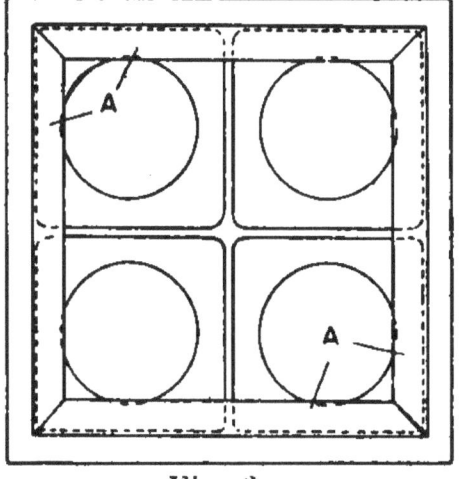

Fig. 6.

close to the sides, for the carbon connections, and one hole in the centre for the connecting socket of the zinc. The wood disc should then be well soaked in molten paraffin wax, and, when cool, well polished by rubbing with a rag. The brass fittings and terminals are fixed to the upper surface of this cover. The brass socket in which the shank of the zinc plate slides should be stout enough and high enough to hold a brass set-screw for the purpose of tightening the shank and holding the zinc plate at any required height in the solution. This socket is to be soldered to a brass strip connecting it with a binding-screw. A semicircular strip of brass connects the two copper strips from the carbon to another binding-screw. These strips are best connected with screws for subsequent easy removal when it becomes

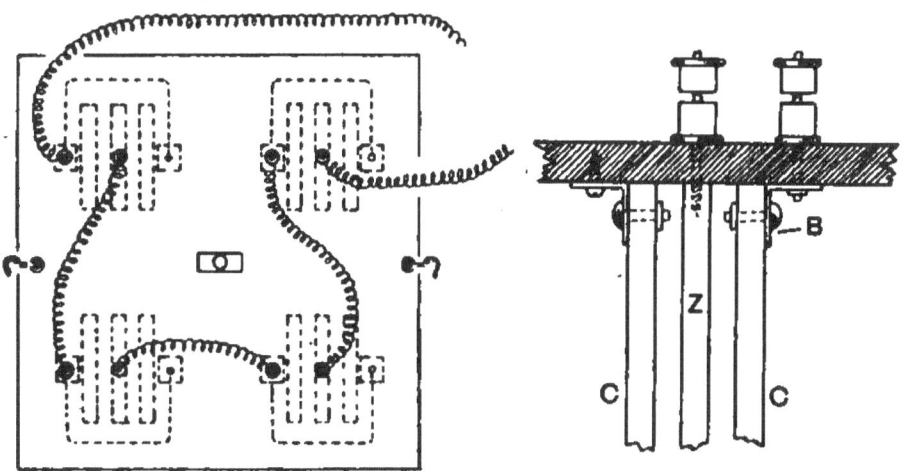

Fig. 7.—Terminal Board
for Portable Battery.

Fig. 8.—Method of Fixing
Carbons and Zincs.

necessary to clean the carbon plates thoroughly or to renew them.

The battery is completed by pouring in the solution made as already described.

Making a Handy Portable Battery.—The bichromate battery, shown in section in Fig. 5 and now about to be described, will be found handy for lighting an 8-volt lamp or working induction coils, motors, etc., and, with care, can easily be carried about when charged. Fig. 5 shows the method of holding the elements out of the liquid while being carried about. When the nut on s is tightened up, the carbons and zincs are held rigid, and the box is made practically water-tight.   Fig. 6 is a plan showing the four glass cells, Fig. 7 is a plan of the terminal board, while Fig. 8 shows the method of fixing the carbons c and zinc z.

The box should be of mahogany not less than $\frac{3}{8}$ in. thick when planed, the inside dimensions being 10 in. by 7 in. by 7 in.   The sides must be notched or dove-tailed into each other, glued and screwed together, and the lid attached with brass hinges and fitted with a brass handle. A small brass plate, with a hole bored in it, should be screwed to the centre of the lid, for the screw s to pass through. A piece of mahogany is next cut about 7 in. square, and made to fit easily within the box ; this is for the terminal board to which the elements are attached.   Four wood strips A (Fig. 6), $\frac{1}{2}$ in. square, are mitred at the ends and screwed to the sides at $\frac{1}{2}$ in. from the top of the box (see Fig. 5) ; care must be taken that they are quite flush with each other on the under side. Brass screws had better be used, three to each strip. These wood strips have indiarubber glued on the under side— a $\frac{1}{2}$ in. wide elastic band, or a piece of bicycle inner tube, is suitable ; this will prevent leakage to any great extent if the battery should be accidentally overturned. The box and the top of the terminal board should now be French-polished, and coated inside with several coats of shellac varnish ; then, if well fitted, it will be quite water-tight.

The glass cells are ordinary Leclanché battery jars, No. 3 size, but any glass jar of square shape, about 5 in. deep, would do. Paper may be wrapped around each, to prevent shaking, if the cells are too loose a fit in the box. Eight carbons, 5 in. by 2 in. by $\frac{1}{4}$ in., are required, and four zincs the same size and not less than $\frac{3}{16}$ in. thick. Eight pieces of brass, $1\frac{1}{4}$ in. by $\frac{5}{8}$ in. by $\frac{1}{16}$ in., are bent at right angles, and drilled for screws and terminals, as shown at B (Fig. 8). Suitable screws, such as are used for sheet metal trunks, can be bought at an iron-monger's; they must be of brass, and have round heads and flat nuts. The terminals are small-size telegraph pattern; and four will require nuts. One terminal is required for each pair of carbons, one carbon being attached with an ordinary wood screw to the terminal board.

The pairs of carbons are connected together with thick insulated copper wire, shown by dotted lines in Fig. 7; the best way of doing this is to solder the ends to the brass angle pieces. The zincs are drilled and tapped to receive the terminal; this is the best method of attaching the zincs, as they can then be easily replaced when used up. A screw-eye is fixed in the centre of the terminal board, and a piece of $\frac{3}{16}$ in. brass wire screwed for a nut or a thumbscrew and bent into a hook, as at s, Fig. 5. The space between the carbons and zincs should not be more than $\frac{1}{4}$ in., two pins being fixed each side of the zinc to prevent it turning and touching the carbons.

When all the screws have been tightened up, all metalwork on the under side of the terminal board must be coated with thick shellac varnish or melted pitch, to prevent the acid fumes from corroding the connections; the tops of the carbons and zincs should be treated in the same way. Two small hooks are shown in Fig. 7, to hold the elements up when the battery is not in use, screw-eyes being fixed in the wood strips A. Four pieces of wood for the terminal board to rest on when lowered should be fixed at the corners of the box, as shown at D (Fig. 5).

The battery cells are shown connected in series, but they can be conveniently connected up in parallel or any other combination.

An alternative but much inferior method of making this battery would be to dispense with the glass jars and simply divide the box into four compartments with wood partitions—each, of course, being perfectly water-tight.

The Fuller Double-fluid Bichromate Battery

The double-fluid bichromate battery has already been referred to and the fact that it maintains a current longer than does the single-fluid form has been noted. The commercial form of the double-fluid cell is illustrated by Figs. 9 and 10, which show an outer jar A of earthenware containing an inner porous pot B in which stands a zinc pole Z with a broad base. The carbon plate C is in the outer jar, and stands in a solution containing 3 oz. of bichromate of potash and 4 oz. of sulphuric acid to the pint of water. The zinc pole stands in dilute sulphuric acid (1 of acid to 20 of water), there being in the bottom of the porous pot an ounce or so of mercury which creeps up the surface of the zinc and maintains the amalgamation.

In good condition the cell described in the previous paragraph has an E.M.F. of 2 volts, the internal resistance varying from 2 to 4 ohms, according to the resistance offered by the porous pot and the condition of the solutions. Commercially, the sizes of the cells vary by pints and quarts. A one-pint size, when the solutions are in good working condition, will give a current of $\frac{1}{2}$ ampere for 20 hours, and a quart size 1 ampere for 20 hours, after which period the solutions must be changed. Smaller currents may be obtained for proportionately longer periods.

In making up the Fuller cell, after inserting the zinc in the inner pot and the carbon plate in the outer jar, fill up both jars with the proper solutions to about the same level. When the zinc and carbon plates are connected, the sulphuric acid passes into the zinc cell and attacks the zinc, forming sulphate of zinc, whilst the hydrogen reduces the bichromate of potash to a lower form. The mercury keeps the zinc well amalgamated, so that hardly any local action takes place. When the solutions become saturated, a secondary action takes place, resulting in the formation of chrome-alum crystals. The formation of these crystals may be prevented by the occasional withdrawal of some of the solutions and replacing with fresh solutions of sulphuric acid and water. The bichromate solution should be of an

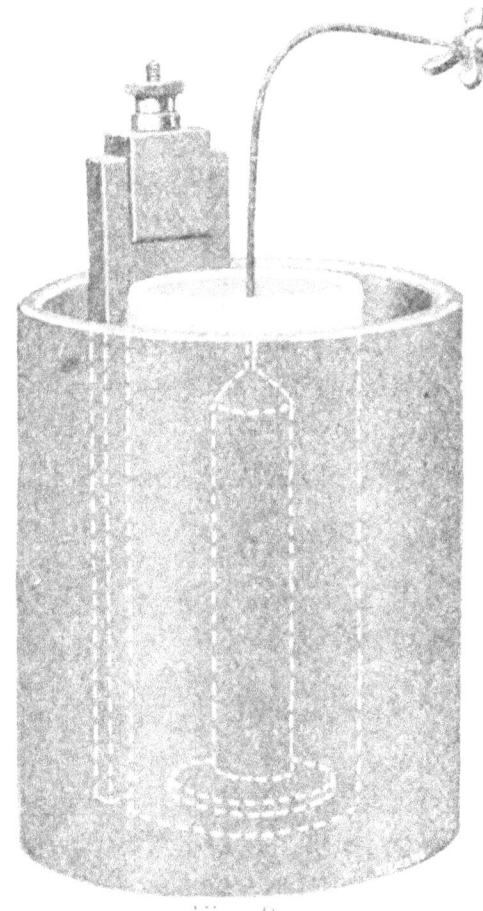

Fig. 9.—Fuller Cell.

Fig. 10.—Section of Fuller Cell.

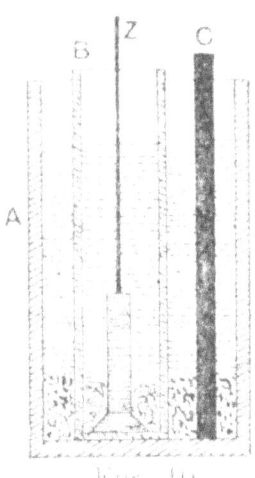

orange colour, but if the solution turns blue, fresh bichromate should be added. If the colour remains orange, and yet the cell fails, withdraw some of the solution from the outer jar and make up with sulphuric acid solution, at the same time also withdrawing half of the solution from the inner pot and replacing with water. If the battery is not required to do much work, it should remain in good condition for three months or longer; but if hard worked, it must be examined oftener.

In cleaning the inner jar of a Fuller battery, place it undisturbed under the water tap until the whole of the old solution is washed away. The mercury can then be used again.

The brass terminals for connecting the cells of a Fuller battery may be purchased ready made at a low price. The heads may be of the butterfly or telegraph pattern, and they are fixed to the carbon plates in lead heads cast on the carbons in suitable moulds. Carbon plates thus capped may also be purchased ready for use. The tops of the carbon plates may be first coated with copper, and this coat tinned as for soldering before the lead cap is put on. This ensures a firmer hold for the lead, and a better connection. For coppering the carbons, *see* pages 20 and 21. The zinc blocks have stout copper wires cast in them, and these are fitted with butterfly terminals soldered to the tops.

The Barrovian Battery

This double-fluid chromic acid battery is adapted for small electric glow-lamps. The porous pots and ebonite case being small, the cells require frequent replenishing. Make two solutions, the first (A), 1 part sulphuric acid to 8 parts water; and the second (B), ¼ lb. of chromic acid in 1 pint of water, to which should be added slowly 1 pint of hydrochloric acid. Be careful to add the acid last. Keep the solutions in stoppered glass or stone bottles in a cool place. To recharge the cells, half fill the porous pots with solution B, then pour about the same quantity of solution A into the case, so that when the porous pots are in position the solutions will be of equal level. The carbons fit inside, and the zincs outside, the porous pots. The E.M.F. is about 2 volts. When not in use, withdraw the carbons and zincs, and rinse with water; also empty the pots and case.

Making Battery Carbons

The production of battery carbons will now be touched upon, inasmuch as amateurs frequently inquire as to whether these can be made at home. In the first place, it must be said that it does not pay to make carbons in small quantities, and that in the absence of special roasting ovens, firing furnaces, and powerful presses only carbons of an inferior grade can be produced. The following outline of the manufacture of battery and arc-lamp carbons at a large works in the United States will show that it is best to leave carbon making to those factories that are especially equipped for it.

The material from which the carbons are made is selected gas retort carbon or petroleum coke, the latter being the solid remaining in the petroleum stills after the oils have been evaporated. It is in the form of irregular chunks of black porous material, and these are ground to pea size and transferred to gas-heated calcining ovens, in which all the volatile matters, etc., are consumed, the residuum being pure carbon. After cooling to a certain degree, the doors are opened and the material is hauled out and spread in even layers over a floor, where it is left to cool.

The cold material is passed through grinding mills and the powdered carbon is separated into different grades in a set of bolting machines, the coarser grades being afterwards re-ground. Next, the material is delivered to steam-heated revolving iron barrels or boxes, in which 20 per cent. of binding material—dried tar—is incorporated with the carbon powder. For the actual shaping of the carbons, two processes are available :—(1) The material is weighed, and placed in moulds, which consist of grooved plates of steel containing from twelve to eighteen forms. The material is carefully packed and adjusted, and then smoothed off with a straight-edge, and the second or upper part of the mould is then pressed upon the powder. The filled moulds are placed on endless chains, which convey them through a gas-heated furnace and then to the head of the vertical plungers of hydraulic presses. After having been subjected to great pressure, they are released and the formed carbons, which are held together by a thin web of material, are removed and placed on a corrugated pan until cool, when they are broken apart by hand and fed one at a time into the strippers, which shave off the portions of the web that adhere. (2) The mixture of powdered carbon and dried tar is hydraulically pressed into compact cylinders, and these are fed, one at a time, into huge presses; these have large cylinders in which are plungers, which force the material through dies, upon the size of which depends, of course, the size of the resultant carbons. The material is forced out into grooved trays and broken off into lengths of about 4 ft. When cool, these are passed through a machine and further cut to the desired lengths.

Pencils produced either by the moulding or the forcing method are baked in the same manner, being carefully piled in the firebrick furnaces in regular rows; a small thickness of carbonising material is placed between each two layers. When the furnace is full it is covered with a kind of clay that vitrifies in the baking process and, covering the bed with a scale, prevents the gas employed as fuel coming in contact with the carbon pencils. The baking lasts for a great number of hours, at the end of which time the top of the oven is removed and the carbons, when cool, lifted out with implements resembling hay-forks.

CHAPTER III

CARBON-ZINC CELLS—THE LECLANCHÉ

POSSIBLY the best known and most extensively used of all primary batteries is the Leclanché, which is made in a large number of shapes, styles, and sizes, the most usual form being that shown in Fig. 11. The square glass jar has at the top an almost circular collar, and contains a cylindrical porous pot in which is placed the negative electrode

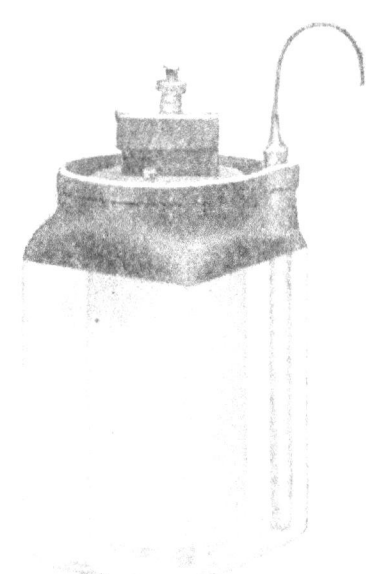

Fig. 11.—Leclanché Cell.

—a carbon plate surrounded by a mixture of granulated peroxide of manganese and carbon. The top of the porous pot is sealed with bitumen, two small holes being left, however, for the escape of gases generated in the pot. The carbon plate has a lead cap, as shown, and to this is attached a binding screw. The positive element is an amalgamated zinc rod, to which is attached an insulated

connecting wire, as illustrated, the zinc being placed
in the jar beside the porous pot. The electrolyte
in the glass jar is a solution of sal-ammoniac
(ammonium chloride) in rainwater. The strength
of the solution should vary with the position of
the battery, as explained later; for a saturated
solution, 14 oz., 7 oz., and 4 oz. of sal-ammoniac
will be required for the respective sizes named
below. Half these quantities will answer for an
ordinary solution. Different makers vary the sizes,
weights, etc., but the following three sizes are
general :—

Size.	E.M.F. approx.	Internal Resistance, approx.	Weight, approx.	Overall Dimensions, including Terminal, approx.
	volts.	ohms.	lb.	in.
3-pint ..	1·60	0·8	5¼	9½ high × 4⅝ sq.
2-pint ..	1·60	1·0	3	8 high × 3⅜ sq.
1-pint ..	1·60	1·4	2¼	7 high × 3 sq.

The advantage of the Leclanché cell is that
there is no local action within it when the circuit is
open and while the cell is not doing work; thus,
for giving intermittent current as for electric bells,
indicators, and certain kinds of telegraphic work,
the cell is durable and economical. Its E.M.F. is good
(1·60) and, especially in the larger sizes, the internal
resistance is fairly low. For telephone circuits it
is found more suitable than other types of cell
generally employed in telegraphy because they have
a comparatively high E.M.F. and a low resistance.
Its disadvantage is that it polarises quickly,
but as in bell work, telephony, etc., it is normally
in use only for a brief period at a time, this disadvantage is not material. Further, there is no waste of
material when the batteries are idle, and they require
but little attention,

When the cells are bought new, it will be noticed that the vent holes at the top of the porous pots are generally sealed up. The usual instructions when first charging are to charge the outer jar, and then, having placed the porous pot inside, to wait until the solution percolates through in sufficient quantity to start the cell working. This takes time, and when the cell is wanted in a hurry, note where the vent holes should be, make two holes in the pitch, and pour a strong solution of ammonium chloride into the porous pot. By making two holes, it can be filled without pausing to let the air out. The outer jar can be charged with the same strength of solution. Use powdered sal-ammoniac.

Before charging, screw the carbon terminal down hard, and give the top of the outer jar, carbon plate, porous pot, and zinc rod a thick coat of hot paraffin wax. This prevents creeping to some extent. The use of a zinc rod one size larger than the cell keeps the wire where it enters the zinc rod well out of the solution. The terminals or wires should not be held with fingers wet with solution. See that the cells of a battery are slightly separated. Avoid spilling solution on the bench, or into the box containing the cells. Always use the largest size cells obtainable, and see that there is no leakage across the line wires, as it soon ruins a battery. Do not have too many cells in one battery, as too much current is allowed to pass and therefore is wasted, not to mention the damage to the electric bells when the circuit is made close to the battery.

Test the cells frequently; reliable voltmeters are cheap. The internal resistance is fairly constant, whether the cell is new or old, providing the external portions of the cell are clean, and therefore the current depends mainly on the voltage. Where the battery is hard worked, be sure to scrape the porous pots and zincs at least once a week. Do not put a handful of sal-ammoniac into the outer jar, thinking

c

to freshen up the cell. The solution is already well charged with zinc chloride, and the sal-ammoniac lies at the bottom.

When the cells are placed in a hot, dry place, charge them with a weak solution (about 3 oz. to 1 qt.) to prevent creeping, so that when the solution has evaporated, the cell can be re-charged instead of adding water.

Cells that have hard, dense porous pots can be improved by drilling a few rows of fine holes into the carbon manganese mixture. Never combine new and old cells to form a battery in parallel, as the electro-motive force of the new cells will soon be reduced to the lower electro-motive force of the old ones. When conditions allow, it is always wise to switch off the battery, so as to obviate the risk of its being run down should the line wires get crossed, or the circuit be accidentally completed in some other way.

The small holes in the top of each porous cell may be fitted with short pieces of tube, which may be of glass or other suitable material; these prevent the holes from being accidentally stopped by dust and dirt. The holes should be kept clear, because the action of the chemicals in the cell produces a gas which should be allowed to escape freely. It is often further advisable, when setting up the cells, to pour a quantity of water or solution of sal-ammoniac through the above-mentioned holes, so as to render the contents of the porous cell thoroughly wet. The chemicals of which the cell is composed cannot work if dry.

The zinc rods in a Leclanché cell become incrusted with a deposit of sal-ammoniac in a saturated solution of this salt, which crystallises out of the solution and creeps up the sides of the rods and of the cell. The remedy in this case is to use only a half-saturated solution; that is, a saturated solution that is diluted to twice its bulk with rainwater, or one made up with one-half the amounts stated in the first paragraph on page 32.

Another cause of incrustation is the employment of a too weak solution. In such cases the rods become incrusted with a deposit of zinc oxide, and the solution appears milky. The obvious remedy is to add sal-ammoniac. Still another cause of incrustation, affecting the sides of the porous cells also, is a leakage in the circuit, thus keeping the battery continually in action. In such cases there is a strong odour of ammonia from the battery. The remedy for this is a discovery and repair of the leakage, a thorough soaking and washing of the cells in warm water, and a fresh sal-ammoniac solution.

The use of impure sal-ammoniac, which by chemical reaction sets free chlorine gas with its characteristic odour of bleaching powder, is the cause of unpleasant smells arising from the Leclanché battery. Wash out the jars thoroughly in clean cold water, and let the porous pots soak for 12 hours in running water if possible. Then make up a fresh solution of sal-ammoniac procured through some really reliable dealers. Make the solution saturated; that is, until it will dissolve no more crystals, and then dilute with clean cold water in equal quantity to the saturated solution.

One method of preventing evaporation and, as a consequence, any unpleasant smell is as follows:—After filling the glass jars to within $1\frac{1}{2}$ in. of the top, pour on a layer of paraffin oil or melted vaseline $\frac{1}{2}$ in. thick. This will prevent evaporation of the liquid and creeping or crystallisation of the salts, but has the disadvantage that it softens the brunswick black and makes an undesirable mess.

For renewing and re-charging Leclanché cells, first throw away all the old solutions; then nearly fill the cells with clean water and pour a wineglassful of spirits of salt into each one; let them stand thus for several hours. (Spirits of salt, muriatic acid, and hydrochloric acid are one and the same thing.) Scrub the porous cells and also the zinc rods with a hard scrubbing brush in hot water, and next stand the porous cells (up to the black line on top) all night in warm water, to which has been added some spirits of salt. In the morning, empty the glass cells, half fill them again with rain-water, and put a handful of crushed sal-ammoniac in each cell. Well scrub the porous pots and zinc

Fig. 12.—Glass Jar for Leclanché Cell.

rods, and clean all the terminals; then put the pots and zincs in the glass cells, and test the battery on an electric bell. If some of the zincs are worn out, replace them with new zinc rods.

If the battery does not work satisfactorily after the above treatment, it is probable that the charge of the porous cells is exhausted; and it will pay better to obtain new cells than to buy materials and recharge them. For a new charge in an old cell is a waste of good material. However, the job may be done in the following manner :—Unseal each cell by melting out the pitch seals in a vessel of boiling water. Well wash the contents in hot water and (if these are to be used again) pick out all the grains of carbon, then throw the rest of the grains away. Soak the cells and the carbon plates in hot water made acid with spirits of salt, and then in clean water for several hours to get all the metal salts out of the pores; next, carefully dry them. Now charge each cell with carbon grains mixed with grains of best binoxide of manganese, in the

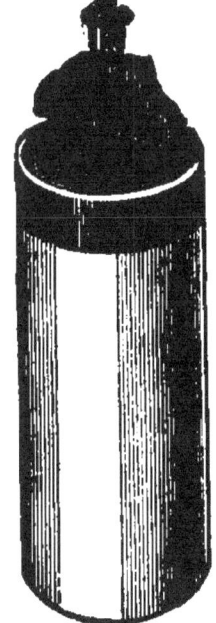

Fig. 13.—Porous Pot Complete.

Fig. 14.—Zinc for Leclanché Cell.

proportion of 85 per cent. manganese to 15 per cent. of carbon, and seal the cells again. But, when recharged, they will not be so good as new cells, and it is hardly likely that they will repay the cost and trouble. Sometimes it will be necessary to soak the recharged cell in a solution of sal-ammoniac, a frequent cause of failure being the dryness of the charge. Another cause may be the choking of the pores of the cell with the products of previous action of the battery. These salts offer a high resistance to the current, and should be dissolved out by

soaking the cells for some hours in warm water
made acid with spirits of salt. The carbon plates
should also be similarly cleansed. In some cases
the salts have crept up under the lead cap of the
carbon and destroyed connection with it, and this
is a cause of failure.

Readers intent on making up their own Leclanché
cells should proceed as follows :—Obtain a glass
jar ; the special shape (Fig. 12) can be bought very
cheaply, the 3-pint size costing only fourpence or
so, and the 1-pint size about twopence. As bought,
the jar has the top of the mouth, both inside and
out, for about an inch down, coated with brunswick
black, paraffin wax, or similar material, that will
prevent the salts formed by the contents from
creeping over the edge. Should a jam jar be used
it will need to be coated ; this is done by thoroughly
cleaning it, heating it, and either dipping the rim
into melted paraffin wax, or brushing on two or
three coats of brunswick black. The porous pot
that goes inside the glass jar contains a carbon
plate with a lead cap, on which is a binding-screw
with connections (*see* Fig. 13). The whole of the
space between the carbon and the pot, to within
$\frac{1}{2}$ in. of the top, should be filled up with a mixture
of equal parts by bulk of crushed coke or retort
carbon and peroxide of manganese, crushed to the
size of very small peas or rice grains, sifted from the
dust and packed in as tightly as possible. So as to
allow the gas formed in working to escape, two
little pieces of glass tube are embedded in the mixture
on each side of the carbon, and then the top should
be sealed up with melted pitch, or pitch and resin
mixed ; the top of one tube is shown in Fig. 13.
The whole of the top of the pot should have two
or three coats of brunswick black, working well over
the lead cap and into the top of the carbon plate,
and down the outside of the top of the pot for about
1 in. Dip the bottom of the porous pot for about

¼ in. into melted paraffin wax, and the negative element is now ready for use. The positive element is generally a rod of drawn zinc (Fig. 14); cast zinc is crystalline and brittle. A hole should be drilled in the top, and a stout piece of guttapercha-covered copper wire either screwed or soldered in. The joint should be well covered with guttapercha or several coats of brunswick black. The zinc should be amalgamated by the following method. With a file remove rough excrescences, etc., and

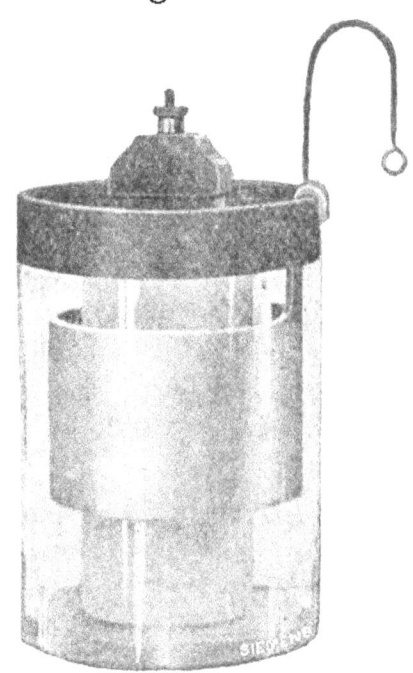

Fig. 15.—Cylindrical-zinc Leclanché Cell.

have ready two glass jars deep enough to take the zincs; one of these should be half full of water containing about a teaspoonful of sulphuric acid, the other a quarter full of the same mixture with a little mercury at the bottom. Dip the rod first in the jar containing the acid and water so as to clean it well, then into the one with the mercury, and by holding it in a slanting position and twisting it round, the mercury can be made to flow all over the zinc. Wipe off the superfluous mercury with a rag, and the zinc rod is ready for use.

It is well to know that should a Leclanché cell be wanted in a hurry and there is no sal-ammoniac at hand, common table salt (sodium chloride) will make a fair substitute.

Modifications of the Leclanché Battery

The commercial modifications of this cell number ten or a dozen at least, and probably individual experimenters could account for very many more.

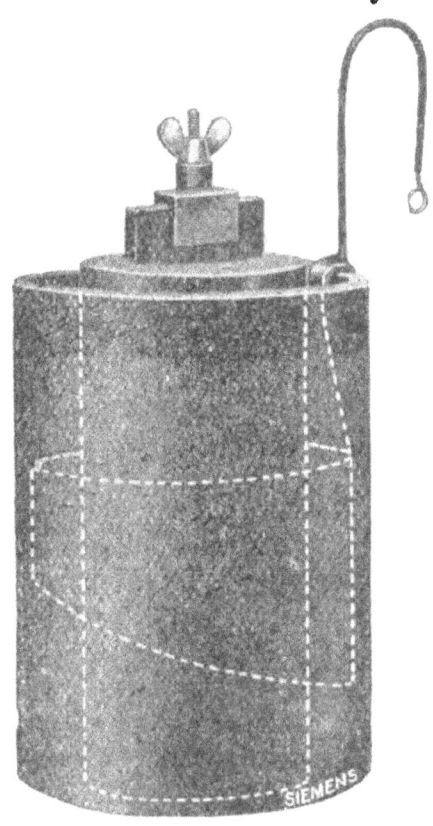

Fig. 16.

Fig. 16.—Cylindrical-zinc Leclanché Cell, Post Office Pattern.

Cylindrical-zinc Leclanché Battery.—Fig. 15 shows a form in which the zinc is a plate bent to cylindrical shape and suspended, in the middle portion of the jar, by means of a substantial lug, which hooks over the rim. To the lug is soldered the wire for making connection, the joint being treated with insulating paint to prevent local action. The customary

porous pot is in the centre of the zinc.   The advantage
gained is a large increase in the active area of the
zinc, it being claimed that the upper and lower
parts of the zinc in the ordinary form of cell are not
acted upon to the same degree as in the central
part. The whole of the surface exposed being
attacked uniformly, there ought to be less waste.
This cell is made in one size only, namely, 6¼ in.

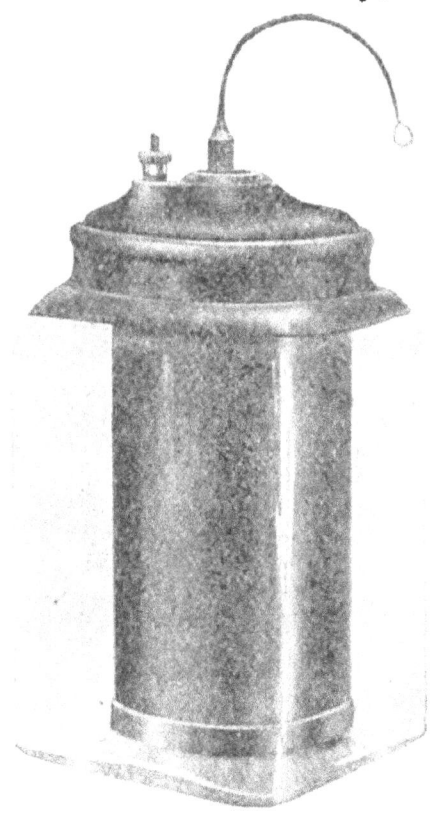

Fig. 17.—Cylindrical-carbon or Carporous Leclanché Cell.

high and 4⅜ in. diameter. Without charge, the cell
weighs about 3¾ lb. The E.M.F. is about 1·60 volts,
and the internal resistance about 0·75 ohm. The
charge is about 5 oz. of sal-ammoniac in water
sufficient to well cover the zinc ; for a saturated
solution 9 oz. of sal-ammoniac would be necessary.
The idea applied in this cell is carried still further
in the British Post Office pattern (Fig. 16), in which
the zinc, bent as before, is cut away, leaving that
portion which, it is claimed, experience has dictated

to be most actively consumed during the working of the cell. The wing-nut form of terminal to the carbon is made of aluminium alloy.

Cylindrical - carbon or Carporous Leclanché Battery.—This has the usual glass container and the inner porous cell, but the latter is surrounded by a perforated cylinder of carbon, the annular space between the cylinder and the porous cell being packed with granulated carbon and peroxide of

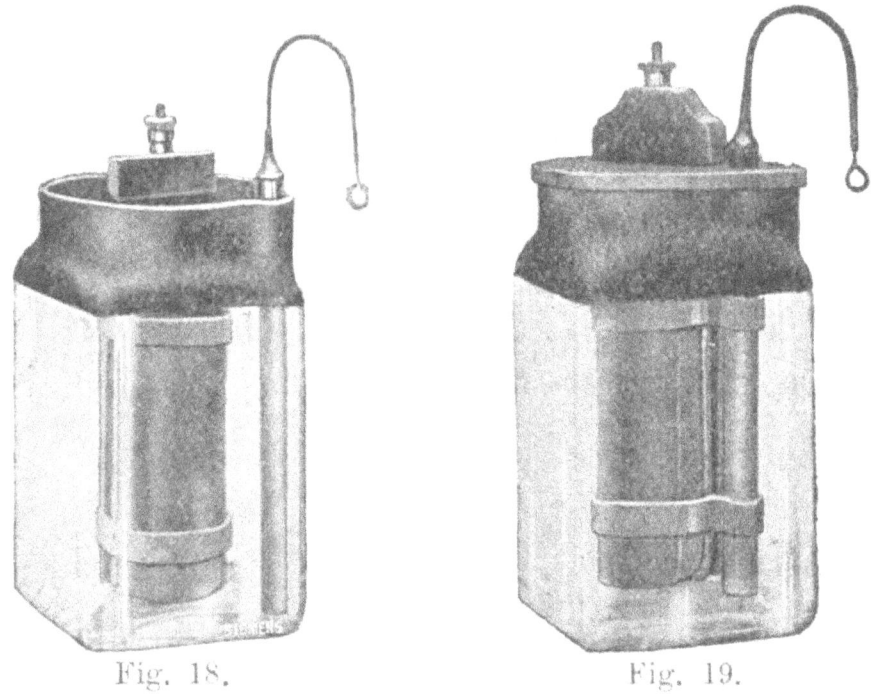

Fig. 18.　　　　　Fig. 19.

Figs. 18 and 19.—Agglomerate Leclanché Cells.

manganese (*see* Fig. 17). The porous cell is brought up higher than the surrounding cylinder of carbon, and the top of the cell is so formed as to serve as a supporting and insulating collar to the zinc rod which forms the positive element. Compared with an ordinary Leclanché cell, the negative element has been greatly enlarged, and the relative positions of the elements reversed. As a result, this form of battery does not so quickly polarise with a heavy current as a battery of the old pattern. The charge is precisely the same as for the ordinary Leclanché.

Agglomerate Leclanché Battery.—In this, the internal resistance has been reduced by doing away with the porous pot and placing the carbon electrode between two agglomerated blocks of peroxide of manganese and carbon, cemented together with gum, and formed under heat and great pressure, the blocks and carbon being held together by simple

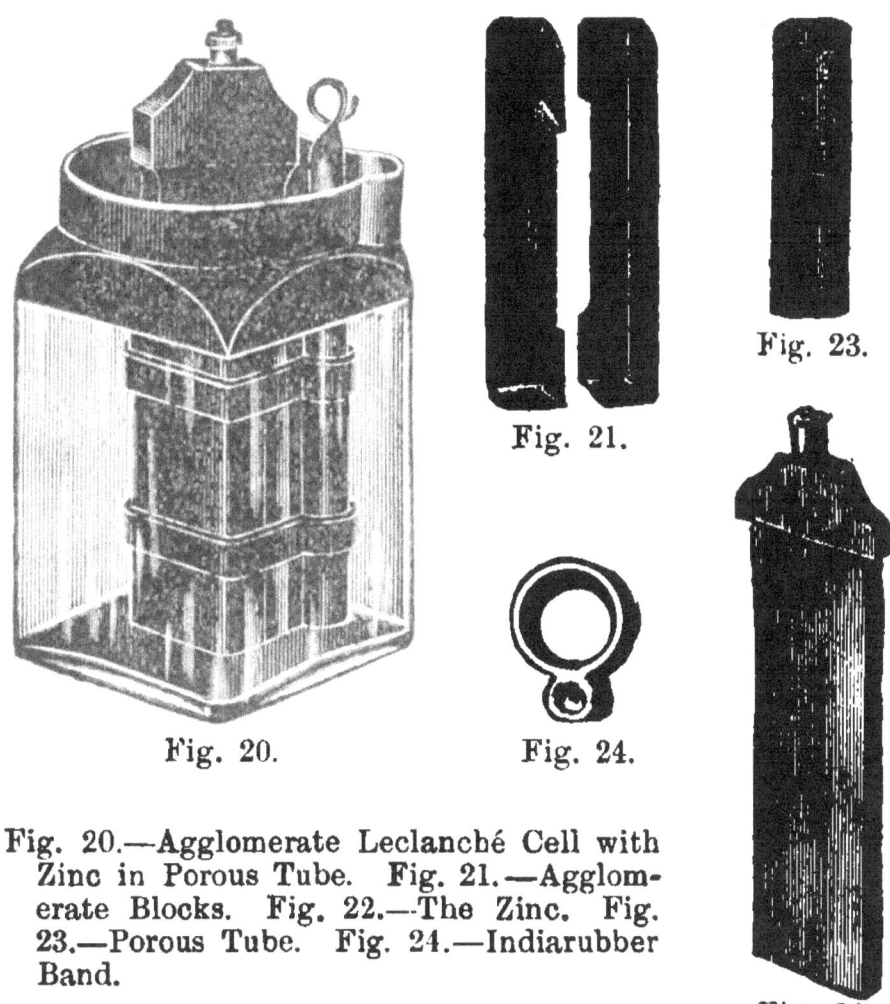

Fig. 21.

Fig. 23.

Fig. 20.

Fig. 24.

Fig. 22.

Fig. 20.—Agglomerate Leclanché Cell with Zinc in Porous Tube. Fig. 21.—Agglomerate Blocks. Fig. 22.—The Zinc. Fig. 23.—Porous Tube. Fig. 24.—Indiarubber Band.

indiarubber bands (as in Fig. 18), or by special bands of the same material formed with an eye for holding the zinc electrode (as in Fig. 19). In another arrangement, the zinc rests against a groove in a porcelain block or gutter, the whole being held together by indiarubber bands. Again, another and common practice is to place the zinc rod in a porous tube, held, as before, by indiarubber bands. The last arrangement is shown in Figs. 20 to 24, of which Fig. 20 shows the complete cell, Fig. 21 the agglomerate blocks, Fig. 22 the capped carbon,

Fig. 25.—Cylindrical-zinc Agglomerate Leclanché Cell.

Fig. 23 the porous tube for the zinc, and Fig. 24 one of the two indiarubber bands with eye.

The table below shows the sizes in which this battery is usually obtainable.

Size.	E.M.F. approx.	Internal Resistance, approx.	Weight, approx.	Overall Dimensions, including Terminal, approx.
	volts.	ohms.	lb.	in.
3–pint ..	1·55	0·50	4¼	9½ high × 4⅝ sq.
2–pint ..	1·55	0·60	2¾	8 high × 3¾ sq.
1–pint ..	1·55	0·70	2	7 high × 3 sq.

The actual composition of the agglomerate blocks may be pyrolusite (the mineral form of manganese peroxide), 40 parts, by weight; carbon, 52 parts; potassium bisulphate, 3 parts; and gum, 5 parts. The potassium bisulphate facilitates the solution of

the zinc salts, which enter the pores. The pyrolusite
or manganese ore used for batteries is not pure
manganese peroxide, but contains some mangano-
manganic oxide (Mn_3O_4) with more or less oxide

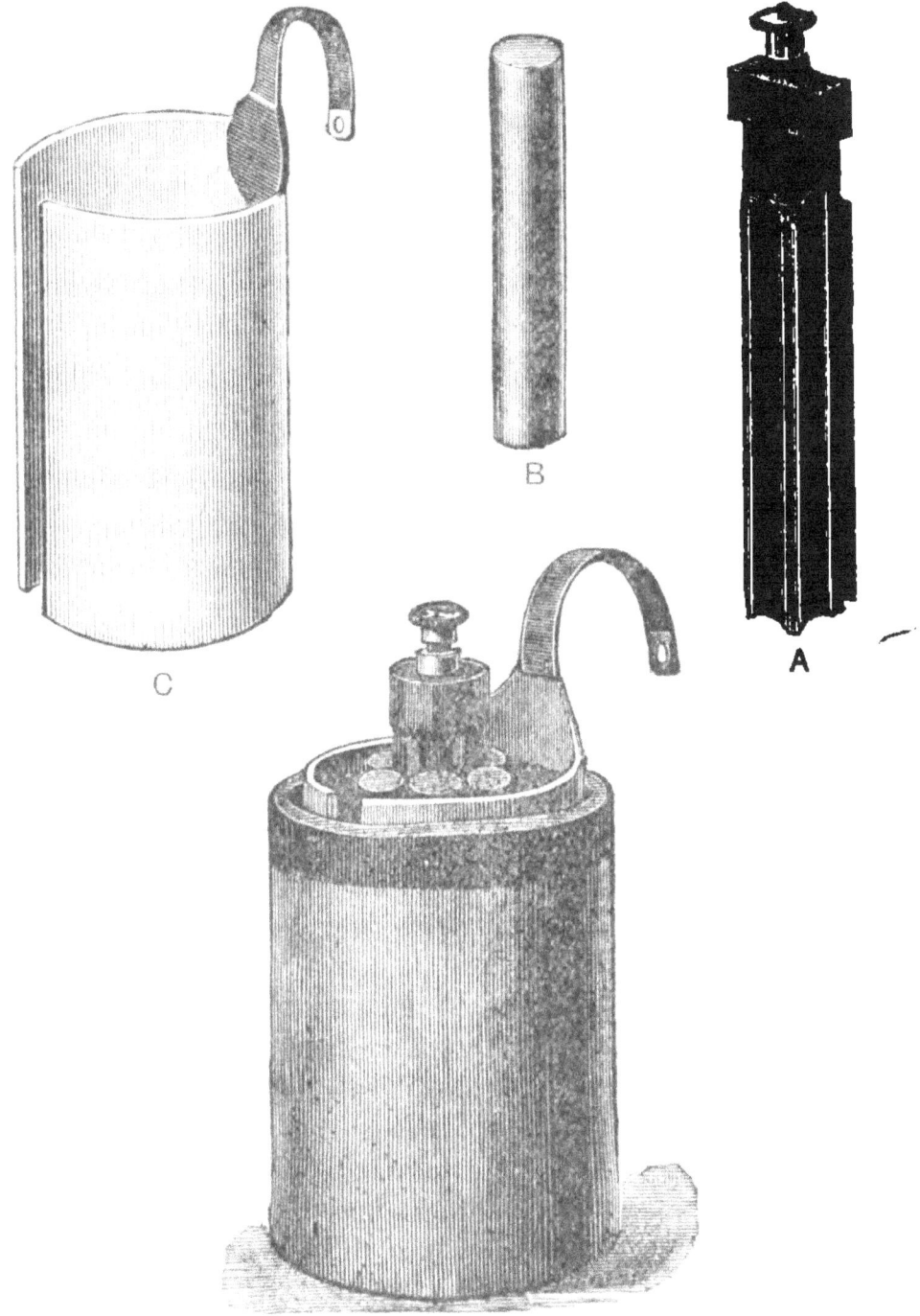

Fig. 26.—Six-block Agglomerate Leclanché Cell and its
Parts.

of iron, lime, and baryta. These impurities reduce the efficiency of the peroxide. In order to convert the mangano-manganic oxide to peroxide, and to dissolve out the other impurities, the ore, broken into small pieces, may be steeped in fairly strong nitric acid for several hours, and then washed free from acid by a running stream of cold water.

The improvement brought about by the agglomerate form is due principally to two facts— the absence of the porous pot lowers the internal resistance, and the compression of the manganese and carbon particles further lowers it, because the mixture is a better conductor than the fluid.

The prevention of evaporation is fairly assured in the case of the battery illustrated by Fig. 19, by the wooden cover which is soaked in melted paraffin wax and then rubbed. In this battery, the carbon plate is formed with a carbon cap, the terminal being fixed into it by means of a special alloy.

Lower internal resistance and a less consumption of zinc are the advantages claimed for the cylindrical-zinc agglomerate battery shown by Fig. 25, as compared with the ordinary pattern of Leclanché. This cell is made in one size only, $6\frac{1}{4}$ in. high and about $4\frac{3}{8}$ in. diameter, the overall height being 8 in., the weight without charge being about $3\frac{1}{2}$ lb., the E.M.F. 1·55 volts, and the internal resistance about 0·3 ohm.

Another special form of the agglomerate Leclanché is the six-block type shown in Fig. 26. Against the advantage that the internal resistance is still further reduced to 0·20 ohm, there is the practical disadvantage that the agglomerate blocks tend slowly to disintegrate and crumble to pieces. The container is an earthenware jar. The carbon electrode is in the centre and takes the form of a star-shaped rod (*see* A), in each of whose six grooves an agglomerate cylinder (*see* B) fits; the whole is wrapped with a strip of coarse canvas held in place with rubber bands. The positive electrode is a zinc plate bent cylindrically (*see* c), the rubber bands preventing it from touching the negative

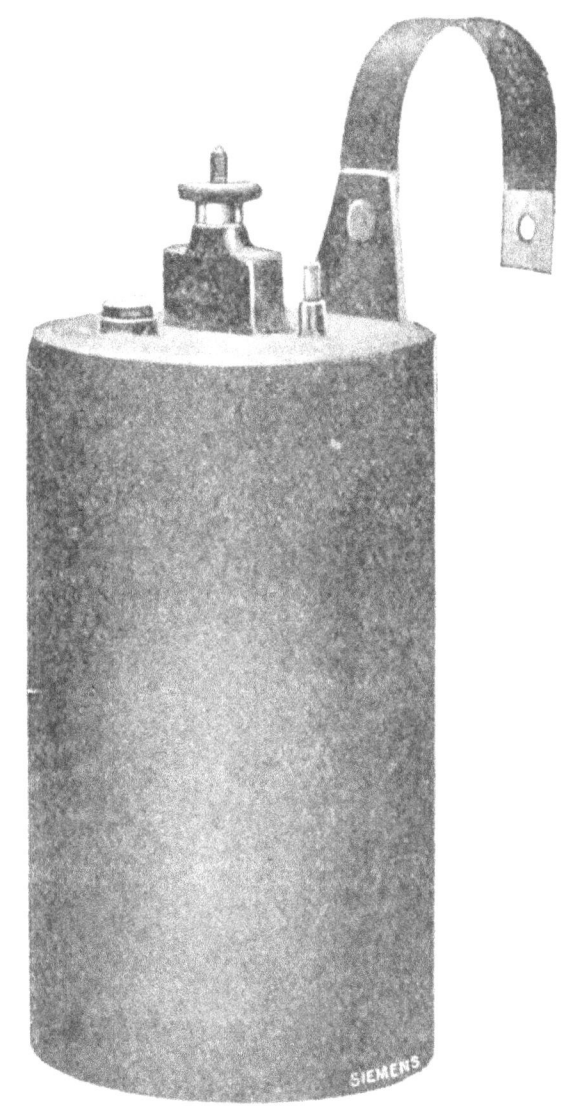

Fig. 27.—Agglomerate Leclanché Firing Cell.

electrode. This cell has a high working capacity, and is well adapted for use on telegraph lines with heavy traffic, or where a single battery serves a large number of bells or indicators.

Two special patterns of Leclanché cells that are extensively used in land and submarine mining operations are illustrated by Figs. 27 and 28, the former being of the agglomerate type, but having a sealed ebonite outer vessel, and the latter, also in a sealed ebonite case, containing a rectangular zinc and

Fig. 28.—Agglomerate Leclanché Firing Cell.

a flat carbon plate with six agglomerate blocks secured by means of rubber bands. The cells are sold charged with sal-ammoniac, so that only water need be added when they are taken into use. Cells that are to be out of use for a considerable time should be emptied and rinsed out with water. When again required, pour in a solution of 1 part of sal-ammoniac in 4 parts of water, filling nearly to the underside of the charging hole and avoiding spilling any on the top of the cell. Batteries of six or ten of these cells permanently connected in series are obtainable in teak cases.

Fig. 29.   Fig. 30.

Figs. 29 and 30.—" Sack " Form of Leclanché Cell.

What is known as the sack form of the agglomerate Leclanché is shown in Figs. 29 and 30. Its shape is suggested by the six-block type, inasmuch as there is a central carbon around which the depolarising mixture of manganese and carbon is moulded, it being kept in place by a wrapping of canvas (*see* Fig. 31). In Fig. 29, the negative element has a porcelain cap, and this, together with the insulating compound at the bottom, keeps it from contact with the cylindrical zinc plate. The overall dimensions are

D

about 9 in. high by 5¼ in. diameter; the weight is about 9½ lb. uncharged, and 10¾ lb. charged; the E.M.F. is about 1·55, and the internal resistance is said to be as low as 0·10 ohm. This cell polarises rapidly, and needs long rests at intervals.

There is not much difficulty in making a sack Leclanché battery at home. The cylindrical sack, measuring from 6 in. to 9 in. long and from 2 in. to 3 in. in diameter, should be made of canvas of

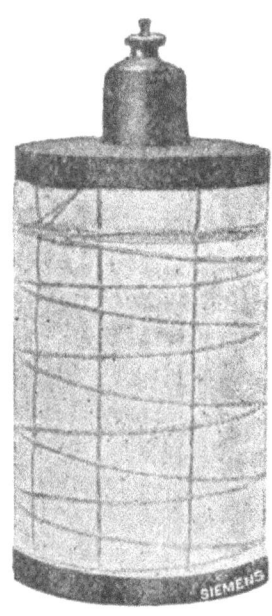

Fig. 31—"Sack" containing Carbon and Depolarising Mixture.

strong but loose texture, avoiding cloth, because this has a high resistance even when wet. The bottom of the sack is a circular piece, and the seams are double-stitched. The carbon must be prepared by attaching a terminal screw to one end by one of the following methods :—(a) Make a mould for the lead head in moulding sand, embedding the terminal screw at the bottom with one tang pointing upwards ; rest the carbon end on this, and fill the moulds with molten lead. When the head is cold, trim off the surplus lead with an old rasp, and coat with brunswick black. (b) Coat one end of the carbon with electrotype copper, as described on page 20, and solder a binding screw terminal to the copper, then coat it with brunswick black as before. The carbon when complete is carefully packed round with the manganese and carbon mixture and inserted into the sack which, after tying at the top in such a way as to leave the terminal free, is ready to be placed in the jar of glass or earthenware. The shape of the sheet zinc element before bending is shown in Fig. 32 ; the zinc is bent into circular form, as already illustrated (*see* Fig. 15, p. 39). Bend the lug over

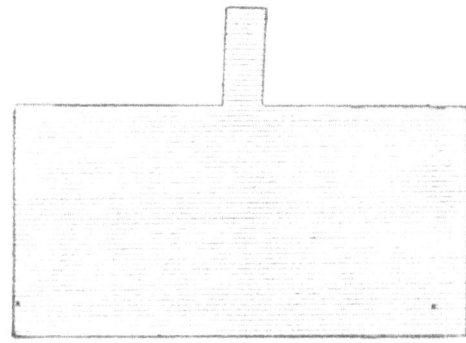

Fig. 32.—Circular Zinc before Bending.

the edge of the jar so that the zinc hangs clear without touching at the bottom ; then remove, and solder a length of stout copper wire to the lug. Amalgamate the zinc by cleaning in dilute sulphuric acid and rubbing mercury over it. There should be a clearance of about $\frac{1}{4}$ in. between the zinc and the sack, and it may be advisable to put thick india-rubber rings round the latter to ensure that there is no contact. The sack and the zinc having been placed in position, pour in the usual solution of sal-ammoniac, and the battery is ready for use.

CHAPTER IV

VARIOUS OTHER CARBON-ZINC CELLS

The Bunsen Battery

THE ordinary form of the well-known Bunsen battery is shown in Fig. 33. A glass or stoneware jar contains a porous pot in which stands a solid, square pole of carbon, to the upper end of which is attached a brass clamp having a terminal screw. A thick cylinder of amalgamated zinc, split as shown, surrounds the porous pot and carries a means of connection with the outer circuit. The porous pot is charged with strong commercial nitric acid, and the outer jar with diluted sulphuric acid.

The E.M.F, varies with the density of the nitric acid used in the porous pot; it is about 1·93 volts when concentrated acid is used, and about 1·89 volts when the acid has a density of about 1·38. For other particulars, *see* the following table :—

Size.	E.M.F. approx.	Internal Resistance, approx.	Weight, approx.	Overall Dimensions.
	volts.	ohms.	lb.	in.
Pint ..	1·93	0·20	3¼	6¼ × 3¾
Quart ..	1·93	0·15	6¼	8½ × 4½

The Bunsen cell gives a fairly constant current, and is suitable for nickel-plating and copper-plating in alkaline solutions, and for gilding and silvering small articles such as chains and trinkets slung on fine wires and offering a high resistance; but for plating large surfaces the Daniell is preferable because this deposits a coat better adapted for burnishing.

The proper way to charge the cell is as follows :—
Pour nitric acid into the porous pot until the level
is less than $\frac{1}{2}$ in. from the top ; for a pint size, about
$4\frac{1}{2}$ oz., and for a quart size about $9\frac{1}{2}$ oz. will be
required. Next fill the outer jar to within $\frac{1}{2}$ in. of

Fig. 33.—Bunsen Cell.

the top with a mixture of sulphuric acid and water,
using for the pint size about $\frac{1}{2}$ oz. and for the quart
size 1 oz. of acid in nine times the quantity of water.

Accidental cracks in the porous jars of Bunsen
cells will allow the acids to mix, and then the current
will diminish. Should any of the nitric acid leak

into the dilute sulphuric acid it will violently attack the zinc and cause failure of current. Failure also results from working the acids too many times before renewing them.

The French Bunsen Battery.—This has the same parts as the ordinary type, but sulphuric acid replaces the nitric acid in the porous pot, with the result that the E.M.F. is lower, being about 1·8 volts, but the polarisation is quicker, the E.M.F. soon falling to 1·6 or 1·5. This type, however, is less troublesome to keep in order, and is free from the noxious fumes which render the use of the ordinary Bunsen almost out of the question in a badly-ventilated room.

The Walker Battery.

The Walker battery, now but rarely used, consists of a zinc positive element and a platinised-carbon negative element standing in dilute sulphuric acid, say, 1 in 10, 1 in 15, or 1 in 20. Both E.M.F. and internal resistance are low, being respectively 0·66 volt and 0·4 ohm.

The Samson Battery.

An advantage claimed for the improved Samson is that the carbon, zinc, and cover are joined so well that it is impossible for a short circuit to occur. The principle of the cell is that of the Leclanché. The carbon and zinc are locked into the cover, which is of a hard material, thus dispensing with the rubber bands formerly employed. To obviate the possible bridging of salts, the carbon and zinc are held ½ in. or so away from the bottom of the jar. The carbon of the improved battery has a fluted lower portion in the form of a hollow cylinder, and a flat top carrying the binding post; these two portions are baked into one piece in the kiln. The top portion is of a different kind of carbon from

the lower. After being baked in the kiln, the top
of the carbon is made red-hot and plunged into
hot paraffin, with which it is impregnated. A
combination of manganese and pea carbon is placed
inside the fluted portion. The close proximity of
this depolariser to the porous carbon increases the
recuperative qualities of the battery. An improvement in the carbon binding-post connections is to
bolt them across the top of the carbon, holding
them in place with a lock nut.

CHAPTER V

COPPER-ZINC CELLS

The Wollaston Battery.

WOLLASTON'S is one of the early and historical batteries that are still in use. It consists of a number of separate cells A (Fig. 34), made of glass or porcelain, each containing a zinc plate zn, centrally adjusted by wooden slips between the halves of a doubled copper plate K, bent round underneath.

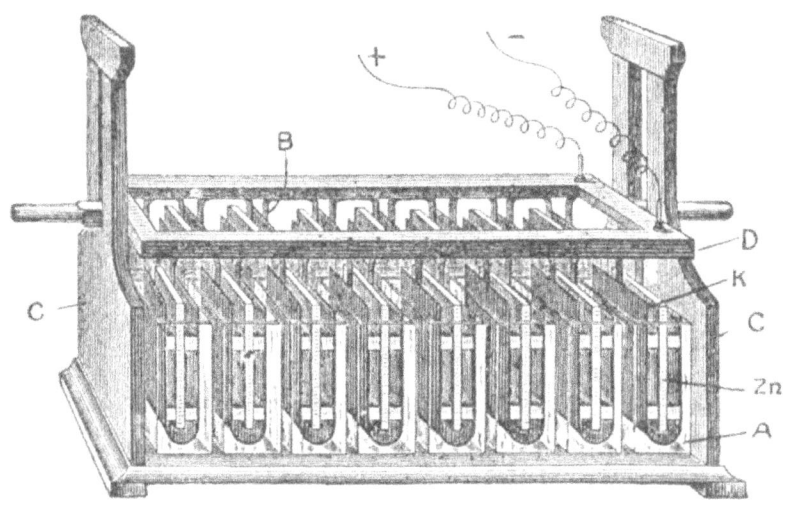

Fig. 34.—Wollaston Battery.

The cells are held in an open box c, side by side in two rows, and all the plates are connected by copper strips B to a frame D, by means of which they can be lifted out of the exciting fluid (dilute sulphuric acid), and the chemical action thus stopped without the need of emptying the cells. This battery is regarded by electro-platers as being the least costly and least troublesome of all, but it is also very inconstant, as its current is apt to fall off rapidly after being set to work; however, it recovers its strength after a few minutes' rest, and it is a

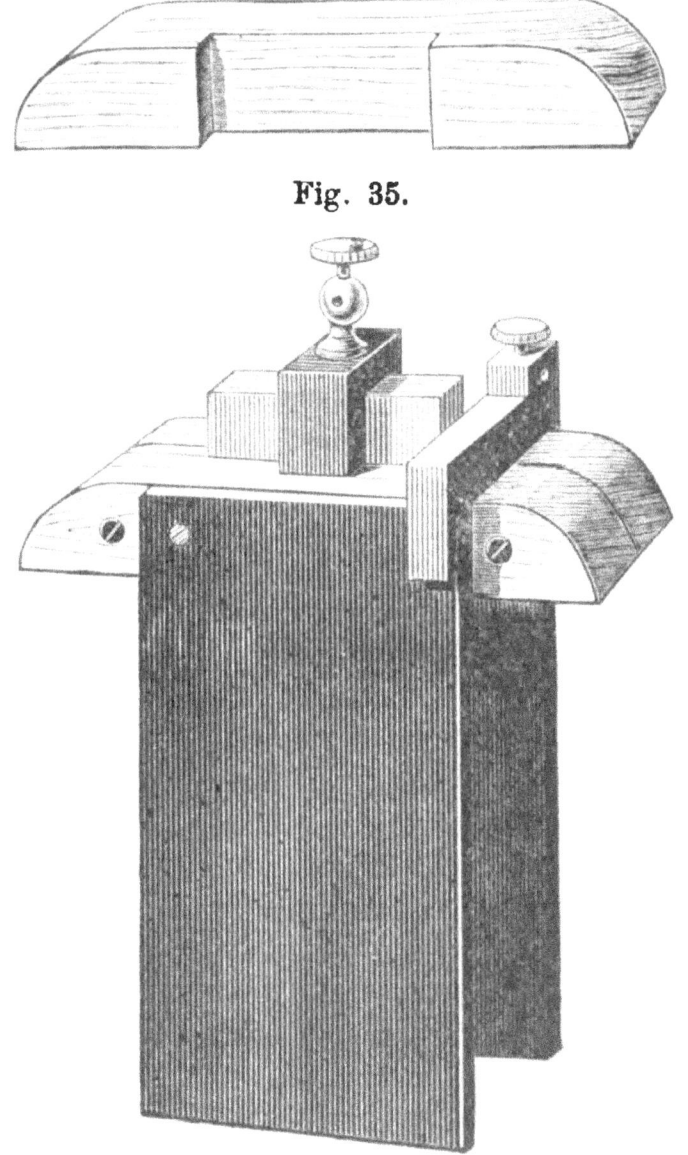

Fig. 35.

Fig. 36.

Figs. 35 and 36.—Parts of Wollaston Cell.

handy battery for short plating jobs and certain kinds of experimental work.

A useful form of the Wollaston battery can be made with three or four jars of glass, stoneware, or porcelain, of any size from 1 quart to 1 gallon,

the larger the better.   Get three or four plates of rolled
zinc, just large enough and long enough to go in the
jars; clean them in hot soda water, rinse in clean
water, and then amalgamate with mercury.   Now
cut out for each cell two hard-wood supports (*see*
Fig. 35), and soak in paraffin wax.   Each zinc plate
is enclosed between two of the supports, the plate
fitting in the recesses, and the supports are secured
to each other by long brass screws passing through
both.   In the top of the zinc is a binding screw (as
shown in Fig. 36).   A pair of copper plates must
now be obtained for each zinc plate, the copper
being slightly larger than the zinc and of any thickness.   They work all the better if they are cross-
scored with a file, or if they have a rough coat of
electrotype copper deposited on them; they work
better still if they are coated with platinum, but
this necessitates the use of a battery and a costly
platinum solution.   The copper plates may be
secured to the wooden supports on each side of the
zincs by brass screws, so short that there is no risk
of their touching the zinc plates; or they may be
clamped with brass clamps (Fig. 36) sold for the
purpose.   When clamps are used, it is always quite
easy to remove the plates for cleaning, and to reverse
the zinc plates so as to wear both ends equally.
The cells are charged with dilute sulphuric acid
(1 in 12).   If the cells are connected up in series
(copper of one cell to zinc of next) a high E.M.F. is
obtained, but when connected up in parallel (all
copper plates together and all zincs together) a
low-tension current of large volume is produced.
The cells may be placed in a wooden tray or in a
shallow box, and all the plate supports may be
secured to a long bar of wood, which is suspended
from an arrangement for lifting all the plates out of
the cells when the battery is not wanted.   This
arrangement will also be found to be most convenient
for controlling the current, as its volume can be

lessened at any time by exposing a smaller surface of the plates to the action of the acid. When the battery is not required for use, the plates should be lifted out of the cells, and if they are not likely to be wanted for a few days, they should be well rinsed in an abundance of water, to free them from acid. It will be necessary to take out the zinc plates occasionally, clean them, and freshly amalgamate their surfaces. This must be done at any time if the plates give off a hissing noise and appear to be blackened by the acid.

Hare's " Deflagrator "

With the special object of obtaining a large surface to both the zinc and the copper plates,

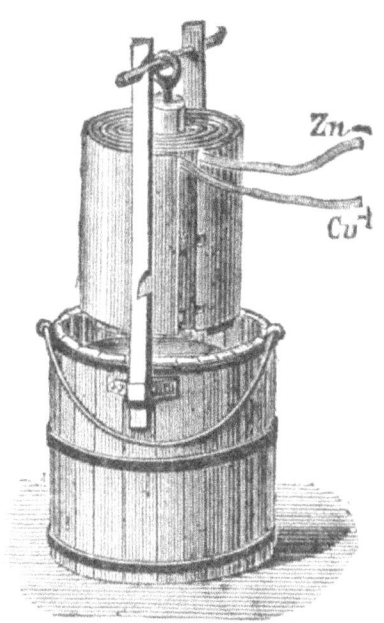

Fig. 37.—Hare's "Deflagrator."

Hare, in another historical arrangement, rolled a copper and zinc plate together to form a double spiral, as shown above in Fig. 37, small strips of wood preventing actual contact. The most noticeable result is a very low internal resistance. It produces a striking heating effect in a low resistance circuit, hence its name.

The Daniell Battery

Prof. Daniell, of King's College, London, invented this very useful battery in the year 1836 with the object of using such a combination of metals and liquids that the chemical effect of the current at the negative plate would not alter the combination, thus avoiding polarisation and maintaining

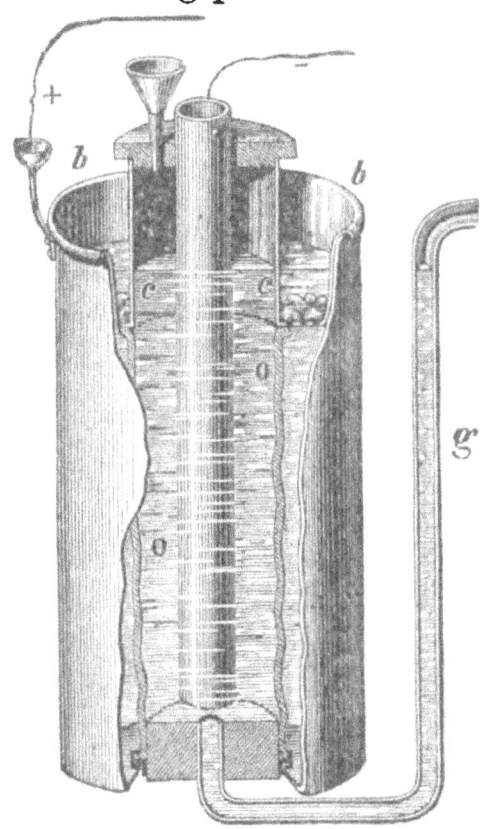

Fig. 38.—Original Form of Daniell's Cell.

an even E.M.F. The negative electrode is of copper surrounded by a saturated solution of copper sulphate, and the chemical effect is that copper is deposited from the solution on the copper electrode, in this way maintaining the combination unchanged. Daniell's arrangement proved effective, and has suggested a large variety of batteries in which, although the details of construction differ, the main principle of obviating the polarisation remains the same.

The original form of Daniell's cell is now of merely historical interest. A copper jar *b* (Fig. 38) contains a saturated solution of sulphate of copper. The jar is the negative electrode and the positive pole. A porous cell *c*, a "membranous tube formed of part of the gullet of an ox," contains a zinc cylinder *o*, the positive electrode, connected to the negative pole. The porous cell *c* contains dilute sulphuric acid, which is poured in through the funnel. The height of liquid in the porous cell is observed by means of a tube *g*, through which superfluous acid is drawn off. To make up for the loss of copper sulphate in the outer liquid, some crystals of this substance are placed in a cage. When the current is flowing, the zinc of the positive electrode dissolves in the dilute sulphuric acid, producing zinc sulphate and hydrogen ; the latter passes through the porous cell and replaces the copper in the copper sulphate solution, the replaced copper being deposited on the copper jar, which becomes thicker. Thus, the liquid in the inner pot becomes zinc sulphate, and that in the outer one becomes sulphuric acid. This, or a corresponding chemical action, is true of all the forms of Daniell cell.

The Daniell in its many forms is still the only constant primary cell, properly so called. Unlike other combinations of dissimilar metals in exciting fluids, the negative plate (copper) does not, in working, become coated with a film of hydrogen gas (causing polarisation and stoppage of the current). Hydrogen is liberated, but no sooner does it appear than it immediately decomposes the copper-sulphate solution into sulphuric acid and metallic copper. The latter is thereupon deposited on the copper element, which thus always presents a clean conducting surface. The sulphuric acid is freed to act on the positive metal (zinc), dissolving it in proportion to the current used. The copper element continually gains in weight by deposited metal.

Thus, Daniell's investigations not only gave to science a constant galvanic cell, but also the whole principle of electro-plating, at one and the same time.

The cell known commercially as the "original Daniell," is that shown in Fig. 39. A round glass jar contains a porous pot inside of which is an amalgamated zinc electrode, and surrounding it a cylindrical copper electrode. The outer vessel is charged with a saturated solution of copper sulphate, and the inner one is charged with sulphate of zinc solution, sulphate of magnesium solution, or dilute sulphuric acid, the liquid standing higher than in the outer jar. The cell is maintained in an active state by adding occasionally a few crystals of copper sulphate to the solution in the outer jar and by withdrawing some of the solution from the inner jar and replacing with distilled or filtered rainwater. When the zinc is used up a new one must be inserted, and when the copper gets too thick it must be replaced by a thinner one. The usual size of shop-bought cell is about 6¼ in. high, 4⅜ in. in diameter, weighs about 3¾ lb. uncharged, has an E.M.F. of about 1·07 volts, and an internal resistance of from 3 to 5 ohms; so it is evident that a high price, in cost of working power, has to be paid for the maintenance of a constant voltage. This type is simple, easily kept in order, and particularly well adapted to closed circuit working; but it should not be kept too long on an open circuit because the copper sulphate then passes through the porous pot to the zinc. When used for constant work on a circuit of high resistance the cell will remain in order wth one charge for several weeks, the length of time being proportional to the size of the cell. When used to give a strong current in a circuit of low resistance, it soon fails owing to the heating and percolation of its liquids. Slight waste occurs on open circuit by reason of diffusion. If not constantly employed in charging, some economy is effected by putting the battery on closed circuit of high resistance (250 to 1,000 ohms). The secondary terminals of a spark coil provide such a path.

For the production of electrotypes, a constant flow of electricity at a low tension is required during a period of several hours. The Daniell battery is one of the best for this purpose, and in making small electrotypes the single cell process is very successful.

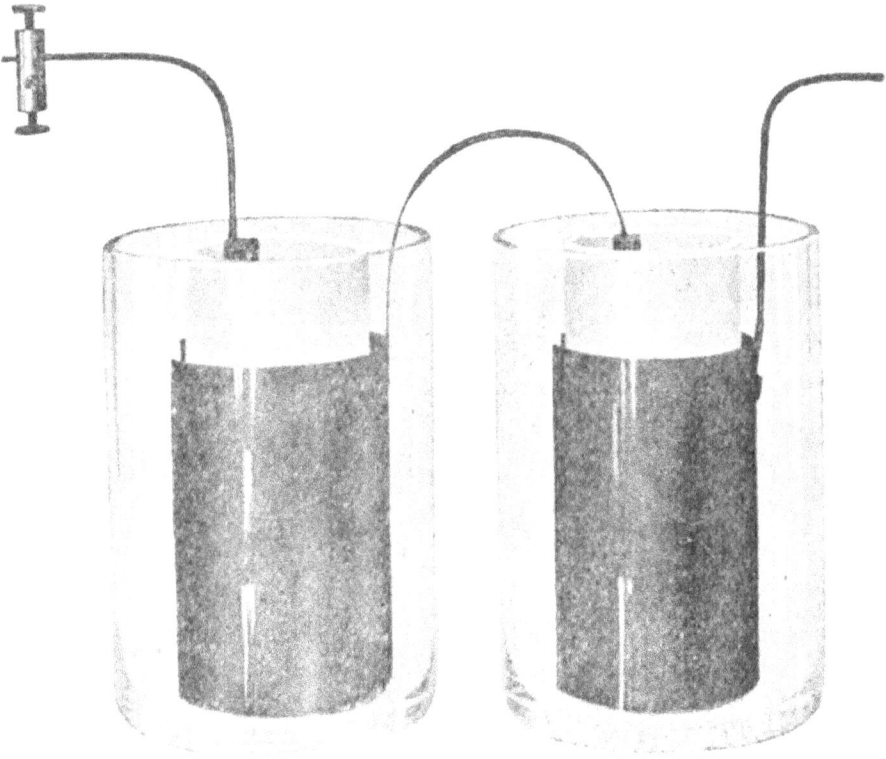

Fig. 39.—A Simple Form of Daniell Battery.

In this process a large glass cell is nearly filled with a saturated solution of copper sulphate in rainwater, and a tall porous cell filled with a solution of 1 part sulphuric acid in 15 parts of water is stood in the centre. In the porous cell is placed a bolt of zinc well coated with mercury, connected to a ring of stout copper wire surrounding the cell. The prepared moulds are suspended by wires to this ring in the copper solution, and as the zinc in the porous cell is dissolved, copper is deposited on the moulds.

Crystals of copper sulphate, for feeding the solution with copper, are suspended in the copper solution in a muslin bag.

The Siemens and Halske Type of Daniell Battery.—The feature of this battery is the employment of a porous diaphragm by means of which the penetration of the copper sulphate through the porous pot to the zinc is avoided, the result being less local action within the cell, increased constancy, and

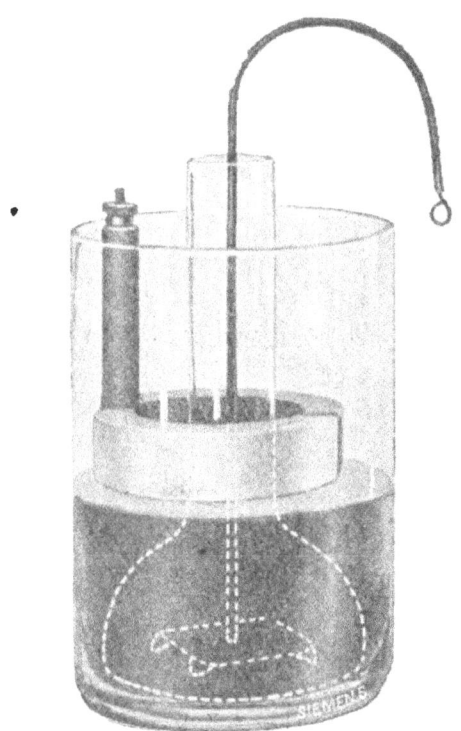

Fig. 40.—Siemens and Halske Type of Daniell Cell.

a greatly increased internal resistance, as compared with the ordinary pattern of Daniell battery. A glass jar contains a bell-shaped porous cup into which is cemented a glass tube (*see* Fig. 40), and within which stands the copper electrode. Surrounding the bell-shaped vessel is a paper pulp diaphragm, upon which rests an amalgamated zinc electrode of annular form with a stalk and binding screw. The purpose of the diaphragm is to prevent the two solutions (a saturated solution of copper sulphate in the inner vessel and dilute sulphuric acid in the outer one) from mixing. This cell is better adapted to open-circuit working than is the ordinary Daniell. It is made in one size, $6\frac{1}{4}$ in. high by $4\frac{3}{8}$ in. diameter; weight without charge, about $3\frac{1}{2}$ lb.; E.M.F., about 1·07 volts; and internal resistance, from 10 to 15 ohms. $3\frac{1}{2}$ oz. of copper sulphate and 1 oz. of sulphuric acid constitute a charge.

This cell is maintained by keeping the inner vessel supplied with crystals of copper sulphate,

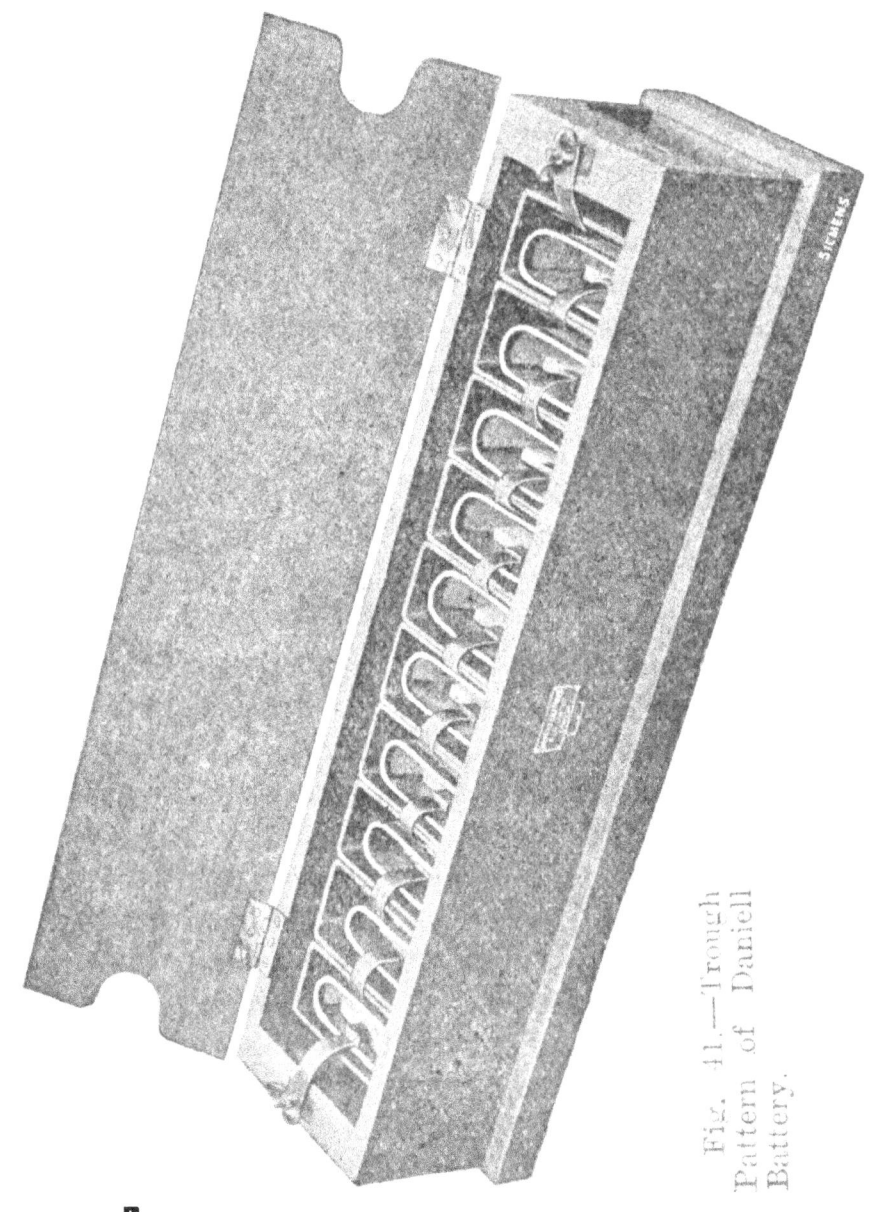

Fig. 41.—Trough Pattern of Daniell Battery.

E

and by occasionally replacing some of the solution in the outer jar with distilled water or filtered rain-water.

The Trough Pattern of Daniell Battery.—A compact and portable form of battery used in telegraphy is that shown in Fig. 41. A teak box, about 30 in. long, by 7 in. square in cross section, holds ten rectangular cells, the containers of which are made of ebonite. The porous pots, whose tops can be seen in the illustration, contain the copper electrodes, there being a zinc plate in the vessel outside the porous cell. The solutions preferred are copper sulphate and zinc sulphate, but magnesium sulphate and dilute sulphuric acid can be used instead. The zinc of one cell is connected by means of a copper strap to the copper electrode of the next, the terminal poles being joined up to brass screws attached to the ends of the box. The E.M.F. of each cell is about 1·07 volts, and of the whole battery about 10·7 volts; the internal resistance is from 2 to 3 ohms and from 20 to 30 ohms respectively. The total weight is 36 lb.

The Meidinger or Balloon Type of Daniell Gravity Battery.—In this and in all other patterns of the " gravity " form of the Daniell cell, the difference between the specific gravities of the two liquids is relied upon to prevent mixing, with the result that there is a big drop in the internal resistance. In the Meidinger cell the outer glass cell is restricted at the bottom (*see* Fig. 42), where there is placed a glass cup containing an electrode, which consists of a cylinder of thin sheet copper having an insulated connecting wire passing out of the cell as shown. The cylindrical zinc electrode is suspended from the rim of the outer vessel and is provided with a terminal wire. The outer jar is charged with dis-tilled water or filtered rainwater to every 5 or 6 parts by weight of which 1 part of magnesium sulphate (Epsom salts) is added. The saturated copper sulphate solution is held in a balloon flask which, as shown, rests upon the edge of the jar and has its neck closed by a cork, through the centre of which is inserted a small glass tube. Care is taken that there is an excess of crystals in the balloon flask. The reason why the cell goes longer without attention

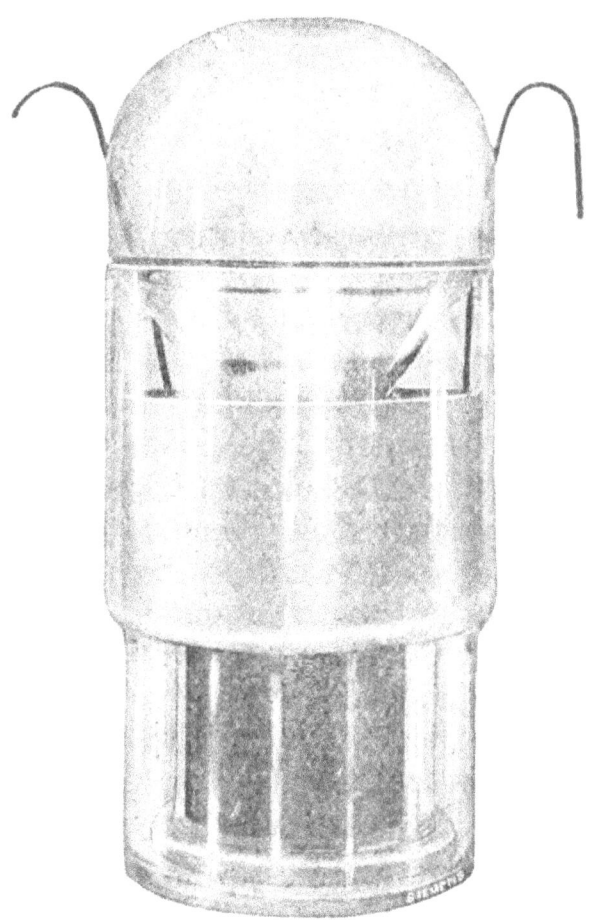

Fig. 42.—Meidinger or Balloon Type of Daniell Gravity Cell.

than the ordinary Daniell is the fact that the balloon flask contains a large quantity of copper sulphate. The cell must not be shaken; otherwise, the two fluids, separated only by their difference of gravity, will immediately mix and destroy the constancy of the current.

  The Meidinger cell is well adapted for 'close-circuit working where the duty is considerable, and for any purpose requiring a moderate but constant current of long duration.

Two sizes of this cell are available.

Size.	E.M.F. approx.	Internal Resistance, approx.	Weight, approx.	Overall Dimensions, approx.
	volts.	ohms.	lb	in.
Line Battery .	1·07	3 to 6	3·56	9¼ high × 4⅝ dia.
Local Battery.	1·07	2 to 4	8	12½ high × 6 dia.

The two sizes take, respectively, about 1½ lb. and 3¾ lb. of copper sulphate, and 2 oz. and 8 oz. of magnesium sulphate.

Important Points in Charging and Operating Daniell Gravity Batteries.—The best results are obtained when the copper element is completely covered with a saturated solution of copper sulphate, and the zinc is immersed in a half-saturated solution of zinc sulphate. There should be a distinct line of separation in the battery liquid between the blue and the white strata, the zinc being suspended in the colourless liquor. The battery, when once set up, should not be disturbed, or the two solutions, separated only by gravity, will mingle, and much waste will ensue. For this reason it is better not to use a ready-made solution of copper to make up the battery in the first instance, but only plain water and crystals. Proceed as follows :—Place the copper plate in position and fill up the jar with filtered rainwater, or with ordinary water that has been boiled at least a quarter of an hour and allowed to cool. A few drops of sulphuric acid may be added to this in case the water is faintly alkaline. Now take a glass tube long enough to reach the bottom of the jar and about 1 in. in diameter (a straight lamp chimney will serve). Down this drop a number of clean blue crystals, previously rinsed in rainwater. Move the tube gently from place to place until the bottom of the jar and the copper plate are evenly covered with a layer of crystals 1 in. or so deep. Use every care not to stir the blue solution as it begins to form and not to cause it to rise up and tint the upper liquid. Place the zinc and cover in position and adjust the level of the electrolyte; set the battery of cells in its permanent position and connect up the elements. If left undisturbed, the battery will begin to yield current in three days, and will attain to its full power in a week, after which it will remain constant for a long period. Examine once a week, and at the first sign of white crystals (of zinc sulphate) forming above the liquid, remove 1 oz. or 2 oz. of the top solution with a glass syringe, and add clean rainwater to make up the original quantity. Add copper crystals when necessary, but use the utmost care not to agitate or mingle the upper and lower liquids. If a half saturated solution of zinc sulphate is used instead of plain water in the first place, the battery will yield current in twelve hours, but this is not recommended; if the cells are allowed to form their own electrolyte their working will be far more steady and reliable.

The Minotto Type of Daniell Gravity Battery.— This is another modification of the Daniell cell, and is made in more than one style. In one, a layer of copper sulphate crystals is put in a glass or stoneware cell; a copper disc, with a guttapercha-covered wire soldered to it, is placed on the layer of crystals, and another layer of crystals is put on the top of the disc. A disc of canvas is then put in and covered with a thick layer of sand or sawdust made wet with a solution of zinc sulphate, and then another disc of canvas is placed on this, with a disc of zinc resting on the whole. The sand or sawdust is employed as a separator in the place of a porous pot. The zinc disc forms the positive element of each cell, and may be coated with mercury to minimise local action. Its connection should be a spiral coil of guttapercha-covered wire, so arranged as to allow the disc to sink freely in the cell as its contents shrink. The cell is charged with a solution of zinc sulphate only.

In another pattern (*see* Fig. 43), there is a copper disc at the bottom of the jar, the guttapercha-covered copper connecting wire passing out at the top of the jar. On the copper disc is a 1½ in. layer of

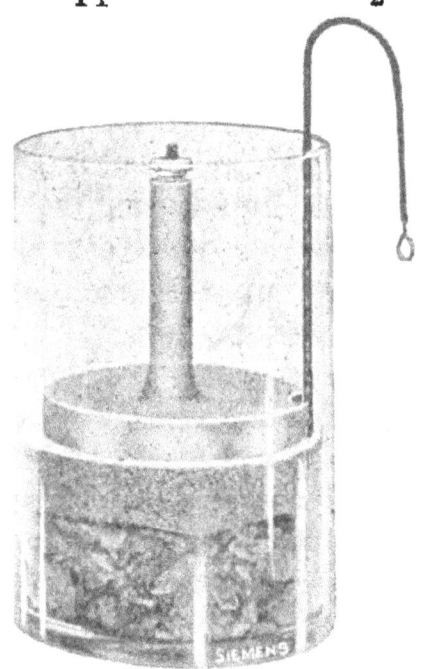

Fig. 43.—Minotto Type of Daniell Gravity Cell.

crystals of copper sulphate ; then comes a disc of blotting paper, cloth or coarse calico, and a 2-in. or 3-in. layer of sand or, preferably, sawdust to act as a separator. Over this is placed a second disc of blotting paper, etc., and finally a zinc disc having a central stalk and terminal screw. The cell is supplied with an exciting liquid merely by pouring in water until the zinc disc is covered.

The Minotto is a very constant form of the copper sulphate series of batteries, and is one of the most suitable to the work of driving the motors of electric clocks and the constant current system of electric alarums. It may also be employed in ringing electric bells, and in doing small electro-gilding jobs and in other electrical experiments where only a small electric current is required. Its internal resistance is higher, and its E.M.F. slightly lower than that of the ordinary Daniell, but it keeps in working order with very little attention for a long time on a closed circuit, where a thin but constant current is required.

To repair such a battery, clean all the elements and connections, and recharge as in making a new cell. The copper and zinc elements may be cleaned in dilute sulphuric acid, and then the zinc discs may be amalgamated with mercury, without first making them smooth.

The following short table shows the two sizes in which the Minotto battery is generally obtainable :—

Size.	E.M.F. approx.	Internal Resistance, approx.	Total Weight, approx.	Dimensions of Glass Jar, approx.
	volts.	ohms.	lb.	in.
Quart ..	1·07	10 to 20	4¼	6¼ high × 4⅜ dia.
Pint ..	1·07	10 to 20	2¼	4½ high × 3¾ dia.

The charge of sulphate of copper is about 12 oz. for the quart size and 8 oz. for the pint size.

Crowfoot, Wheel and Star Types of Daniell Gravity Battery.—The names of these three patterns are suggested by the forms taken by the zincs (*see* Figs. 44 to 46).

In the first, the crowfoot zinc is hung on the rim of the glass jar (*see* Fig. 44), the copper element consisting of three copper plates riveted together in the centre, an insulated wire being joined to it.

In the second, a cast circular zinc, suggesting a wheel, is suspended from the wooden cover by means of a screwed brass rod, which projects and forms the negative terminal, the other electrode being a dished copper disc resting on the bottom of the jar and having an insulated connecting wire projecting through the cover, as in Fig. 45.

In the third, the star-shaped zinc is suspended by a copper wire from a brass tripod resting on the

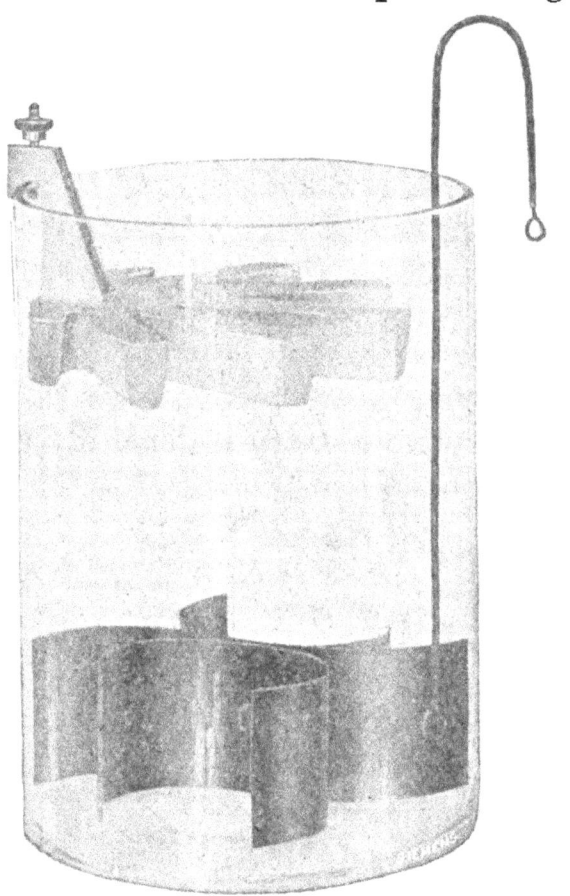

Fig. 44.—Crowfoot Type of Daniell Gravity Cell.

rim (*see* Fig. 46), the other electrode being practically the same as that in the crowfoot pattern (Fig. 44).

All three patterns are charged in the same way. A solution of 1 part by weight of sulphate of zinc in 10 parts of water is poured into the jar, and then crystals of copper sulphate are dropped into the liquid until they reach a level of about 1½ in. from the bottom. The liquid should be on a level with the top of the zinc.

For the two sizes of the crowfoot pattern, 2 lb. and ¾ lb. of copper sulphate, and 9 oz. and 3½ oz. of

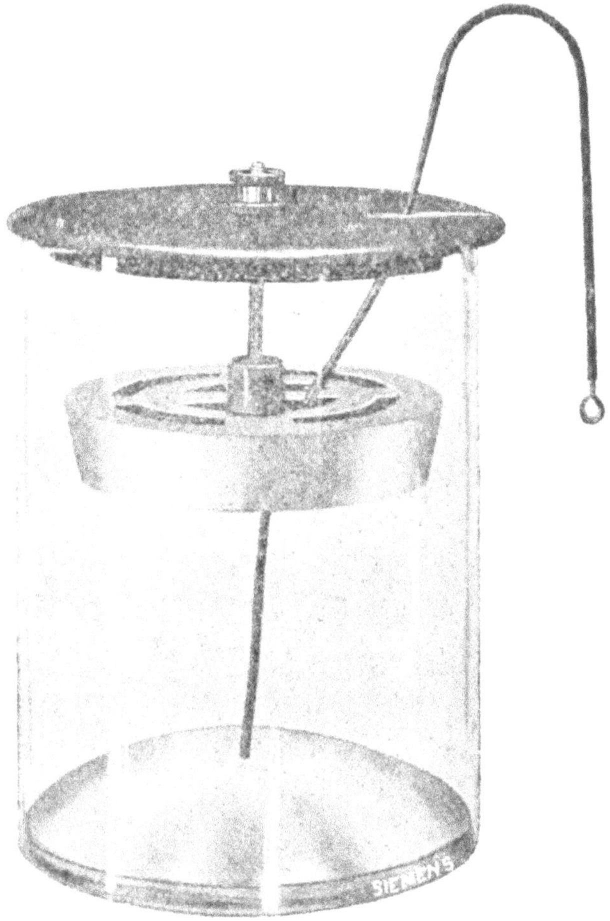

Fig. 45.—Wheel Type of Daniell Gravity Cell.

zinc sulphate will be required respectively. Similarly, for the wheel pattern, 2¼ lb. and 8 oz. of copper

Name of Cell.	*E.M.F. approx.*	*Internal Resistance, approx.*	*Weight, approx.*	*Dimensions of Glass Jar, approx.*
	volts	ohms.	lb.	in.
Crowfoot Pattern	1·07	1½ to 2	6·8	8½ high × 6⅜ dia.
Do.	1·07	2 to 4	2·3	6¼ high × 4⅜ dia.
Wheel ..	1·07	2 to 2½	6·7	8½ high × 6⅜ dia.
Do.	1·07	2 to 4	3·4	6¼ high × 4⅜ dia.
Star ..	1·07	1 to 2	7·62	8¾ high × 6¾ dia.

sulphate, and 9 oz. and 3 oz. of zinc sulphate will
be necessary.   For the one size of the star pattern,
use about 2½ lb. of the one and 10 oz. of the other.

Dimensions of these three patterns of Daniell
cells are given in the table at the foot of page 73.

Fig. 46.—Star Type of Daniell Gravity Cell.

The Callaud Type of Daniell Gravity Battery.—
This is a modification by Callaud of the Daniell
battery, and therefore should bear his name. As
first introduced, it had a circular plate of copper
at the bottom of each cell, covered with a layer
of sulphate of copper crystals, over which there
was poured a solution of sulphate of zinc. The
zinc element was suspended above the copper element

in this solution. It was introduced as a telegraph battery, and used for this purpose in France and America. It is one of the best for this purpose on long telegraph lines of high resistance, where the call for a small volume of current is so frequent as to be nearly constant; but it is altogether unsuitable to work demanding a large volume of current through a low resistance, as in charging accumulators. The E.M.F. of each cell is the same as that of the Daniell (a fraction over 1 volt); but the internal resistance may be anything from 2 ohms to 5 ohms, therefore the available amperage for outside work is exceedingly small, say, from 0·2 to 0·5 ampere each cell.

The present day form of the Callaud cell is shown by Fig. 47. From the rim of the glass jar is suspended, by means of three copper

Fig. 47.—Callaud Type of Daniell Gravity Cell.

hooks, a cylindrical zinc plate, the other electrode being a sheet copper cylinder c, as shown, riveted to a stout copper wire D (insulated with guttapercha), soldered and riveted to a zinc pole of the adjoining cell A of the battery. Thus each cell has two cylinders, zinc and copper. Pour into the glass jar some water containing a tenth part of a saturated solution of sulphate of zinc. Next drop into the liquid crystals of copper sulphate until they cover the copper cylinder; or instead, by means of a tube, gently convey a saturated solution of copper sulphate to the bottom of the jar. Make up the liquid if necessary until it is about ¼ in. from the rim. Carefully avoid any shaking liable to cause the liquids to mix.

Callaud cells are made in the following sizes :—

Size.	E.M.F. approx.	Internal Resistance, approx.	Weight uncharged, approx.	Dimensions of Glass Jar, approx.
	volts.	ohms.	lb.	in.
1-Quart .	1·07	2 —3½	2¾	6¼ high × 4⅜ dia.
2-Quart .	1·07	1½—2¼	4	9½ high × 5 dia.

Automatic Filler for Daniell Gravity Cells.—In Fig. 48 is shown a device that will automatically

Fig. 48.—Automatic Filler for Daniell Cell.

Fig. 49.—Wooden Holder for Filler.

fill a gravity cell, and keep the level of the liquid at almost the same point at all times. It invites comparison with the balloon cell described on p. 66. A wooden holder, as shown in Fig. 49, is placed across the top of each cell and the neck of an inverted bottle containing liquid is inserted in the hole. As soon as the level of the liquid in the jar is lowered below the mouth of the bottle by evaporation, the liquid will run from the bottle and supply the deficiency.

CHAPTER VI.

COPPER-OXIDE ZINC CELLS

The Edison-Lalande, or Lalande-Chaperon Battery

ONE of the best-known forms of copper-oxide cell is the Edison-Lalande, in which the elements are respectively zinc, and a compound plate of compressed black oxide of copper sandwiched in between two metallic copper elements, which grip the oxide plate at its edges and leave the face exposed (*see* Fig. 50). The container is of porcelain

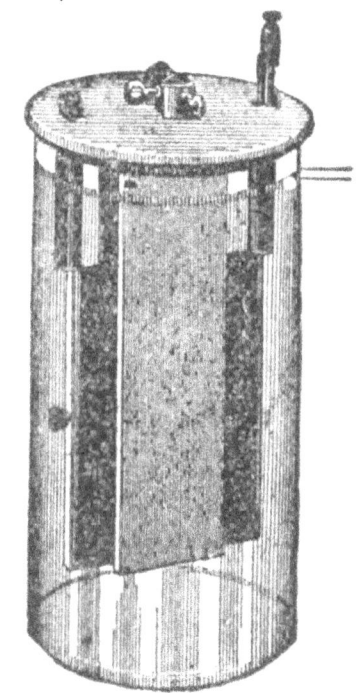

Fig. 50.—Edison-Lalande or Lalande-Chaperon Cell.

or glass with a cover of porcelain. The exciting solution is caustic potash, and sometimes a layer of paraffin oil is poured over the top to exclude air and prevent the absorption of carbonic acid. The E.M.F. is low, about 0·75 volt per cell; but this is compensated for by the low internal resistance and depolarising properties of the cell. A cell 7 in. in diameter and 18 in. high, will give a steady current of 1 ampere continuously for 600 hours.

This battery is excellent for central-battery telephone working, closed circuit alarm systems, and for other purposes requiring a constant current.

In explanation of the principle of this class of battery, it should be said that when zinc immersed in a solution of caustic soda is brought into contact with a more negative element such as copper, the zinc dissolves as oxide in the alkaline liquid, forming a weak compound with the soda (sodium zincate, Na_2ZnO_2), the solution of which, being heavier than the soda, falls to the bottom of the cell, while hydrogen is given off from the copper. In the above cell, in which oxide of copper is used as the cathode, the hydrogen reduces the oxide to metal, and the current is continuous till the whole of the oxide of copper is reduced. The layer of oil on top of the caustic soda solution prevents access of atmospheric air, the carbonic acid in which would convert the soda into carbonate, and soon render the cell quite useless.

As the black oxide of copper, forming the negative element, parts with some of its oxygen in the process of polarisation, this must be renewed when its store of oxygen has become exhausted. At the same time the caustic alkali solution should be renewed. It may be possible partially to restore the activity of the copper oxide by heating it to redness; but the cost of new plates will not be high, and these will be more satisfactory.

It is not necessary to amalgamate the zinc plates with mercury. Thick plates will last a long time, the rate of consumption being in a direct ratio to the current obtained from the cell, but the electrodes must be always kept under the coat of oil on the surface of the solution.

Making an Edison-Lalande or Lalande-Chaperon Cell for Accumulator Charging.—A modification of the Lalande-Chaperon cell suitable for charging accumulators can be made without difficulty. Obtain four glass jars 6¾ in. high and 3½ in. in diameter, and fit each with a wooden cover, so formed as to have one part loosely fitting the jar inside, and the other part resting on its edges (*see* the section, Fig. 51). Soak the covers in melted paraffin wax to render the wood waterproof and non-conducting. In the centre of each cover fit a brass

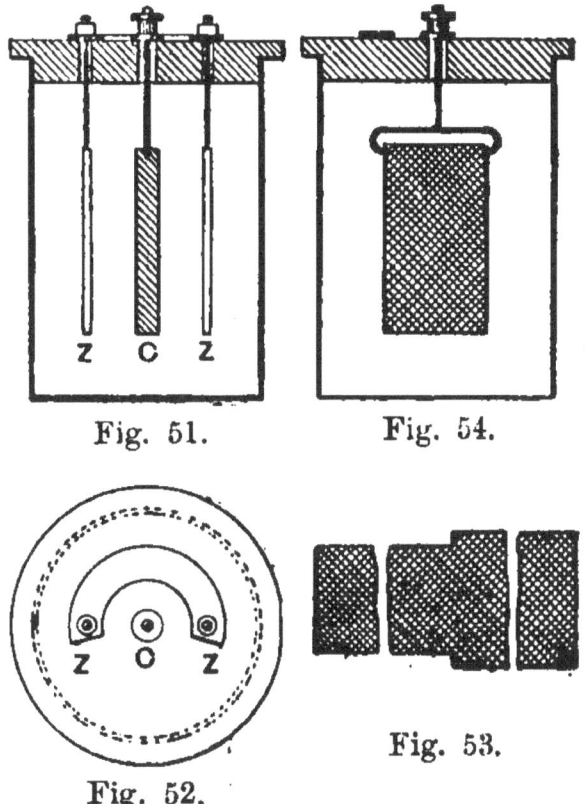

Fig. 51.

Fig. 54.

Fig. 52.

Fig. 53.

Figs. 51 to 53.—Vertical Sections and Plan of Edison-Lalande Cell. Fig. 54.—Copper Gauze for Edison-Lalande Cell.

terminal, for the copper-oxide element; and a piece of brass (*see* Fig. 52), fitted with another terminal, for the two zinc elements. The negative element is prepared from copper gauze having about thirty-two meshes to the inch, formed into envelopes 4 in. by 3 in. by ⅛ in., and enclosing a paste consisting of commercial black oxide of copper and plaster-of-Paris. The strips of copper gauze should be 9 in. long and $3\frac{3}{4}$ in. wide, cut to the shape shown at Fig. 53. Each strip must be coated to a thickness of $\frac{1}{8}$ in. with a paste made of 9 parts of black oxide of copper and 1 part of plaster-of-Paris, with enough water to give a stiff consistency. The end of the gauze is then turned up, and the sides folded in to form the envelope, and the whole put between two boards and heavily weighted until quite hard. The plates must then be fitted to wire connections. Obtain four 10-in. lengths of No. 16 copper wire, lay them across the copper gauze at the top of each plate, and fold the gauze down tight on the wire; then bend the ends over the top, and twist them tightly together to form a tang for connecting the plate to the brass terminal above (see Fig. 54). Next get eight pieces of sheet zinc, 5 in. by $2\frac{1}{2}$ in. by $\frac{1}{16}$ in., and fit and solder to their tops a sufficient length of No. 14 brass wire to connect them with the semicircular brass terminal on the cover, having first cut a screw thread on each wire, and fitted small nuts to each for the purposes of securing electrical connection. Eight rolled zincs, 5 in. high and any convenient diameter, would give better service than the sheet zinc positives. All the elements should be so connected as to be easily removed for cleaning and renewal when required.

Each cell made as above described should be charged with a solution of caustic soda, in the proportion of $7\frac{1}{2}$ oz. of caustic soda and 1 dram of hyposulphite of soda in 3 pints of water. The solution should just cover the copper oxide plate in each cell, and the surface of the solution should be coated with $\frac{1}{2}$ in. of vaseline or oil, to exclude air and prevent evaporation.

These four cells, connected in series (positive of one cell to negative of other), will yield a current of 2 amperes at an E.M.F. of 2·8 volts for twenty-five hours with one charge, and may be left charged ready for use for periods of several weeks.

The Neotherm Battery

This was introduced into England by Siemens Bros., Ltd., and it differs from other types of cells (*see* Figs. 55 and 56). It is a copper-oxide battery containing a caustic soda electrolyte. The cell has the advantage that the depolariser or copper oxide can be regenerated over and over again simply by heating the rectangular cast-iron containing vessel, which is lined internally with the copper-oxide depolariser. Inside this vessel is placed a zinc plate. There is an enamelled iron cover, which screws down liquid-tight on a rubber packing gland, and there are two terminals, one connecting to the iron case and the other to the zinc. When the cell is in action the copper-oxide lining is gradually converted into red spongy copper, practically chemically pure copper. By the simple act of heating the iron case in a clear fire or muffle, this pure copper is converted back into copper oxide. The cell has then simply to be refilled with fresh liquid, and it is ready for a long period of use. The E.M.F. per cell may reach 1·1 volt, and it is possible to discharge at a rate of from 1 ampere to $2\frac{1}{2}$ amperes or even more. The following table shows the sizes in which this cell is made :—

Size.	E.M.F.	Internal Resistance.	Weight.	Dimensions.	Capacity.
No.	volts.	ohms.	lb.	in.	amp. hrs.
1	1·0 to 1·1	—	12	$8\frac{1}{2}$ high $\times 8\frac{1}{4} \times 2\frac{1}{4}$	150 at 1 amp., and 100 at $2\frac{1}{2}$ amps.
2	1·0 to 1·1	—	27·7	$13\frac{1}{4}$ high $\times 9\frac{1}{2} \times 2\frac{1}{2}$	300 at 2 amp., and 200 at 5 amps.

The following instructions on setting up, working and regenerating the Neotherm cells are given on the authority of the makers. The requisite quantity of caustic soda for a first charge is supplied with the cells. It is contained in sealed lever top tins, each holding sufficient for one cell. When, for subsequent re-charging, it becomes necessary to weigh out the soda from bulk, from 9 to 11 oz. must be taken for each No. 1 cell, and about 24 oz. for each No. 2 cell. Casks, etc., containing the soda must be kept well closed. The above quantities

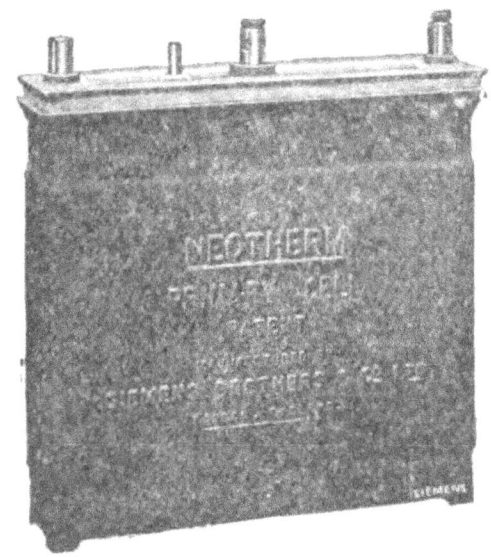

Fig. 55.—Neotherm Cell.

of soda should be dissolved in about 35 oz. (1¾ pints) of pure water for each No. 1 cell, and about 84 oz. (4¼ pints) for each No. 2 cell. When setting up a number of cells at one time, the soda should be dissolved in a separate iron vessel, which should not be rusty. The liquid should be well stirred until all the soda is dissolved; as the soda dissolves, the liquid becomes hot, and it must be allowed to cool before it is used. The density or specific gravity, as measured with a hydrometer, should be 1·220 when the solution is warm, or 1·241 when cold. If the reading of the hydrometer is too low, more soda should be added, and if too high, the quantity of water should be increased.

When new Neotherm cells are to be charged, proceed as follows :—The copper screws on the right-hand and left-hand sides of the cell should be removed. The cover, with the zinc plate attached, can then be lifted ; avoid interchanging the covers when replacing. Remove the wood shavings which have served as a packing for the zinc plate, and proceed to pour in solution until its level lies nearly to the under side of the lugs which receive the screws

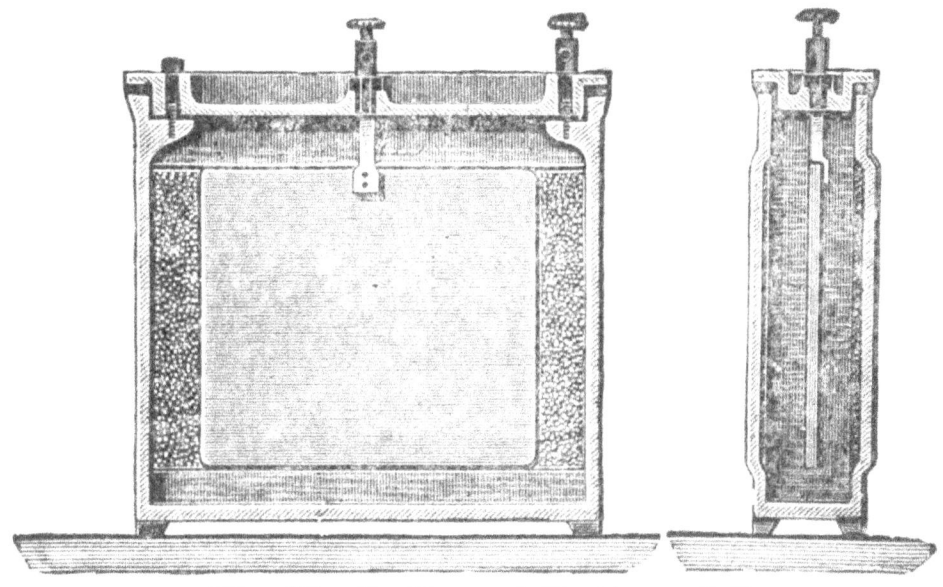

Fig. 56.—Sections of Neotherm Cell.

of the cover. The upper edge of the zinc plate, when in position, must be completely covered by the solution, as, otherwise, an unequal consumption of zinc will ensue. If, when the solution has been made up in a separate vessel and a quantity of solution has been poured into any cell, it is found that the zinc is not completely immersed, more solution must be added. It will not do to add plain water. If an iron vessel is not available, and it is necessary to dissolve the caustic soda in the cell itself, 11 oz. should be placed in each No. 1 cell, and 24 oz. in each No. 2 cell. Water should be added so as to reach nearly to the underside of the lugs which serve to fix the cover. The liquid must be stirred vigorously, and, if possible, the density measured. If the amount of solution in the cell is not sufficient to completely cover the zinc plate, plain water must be added to make up the necessary quantity. The zinc must not be placed in the cell until the solution is cold. Before the zincs are placed finally in the cells, the solution should be allowed to stand in the cells for about half an hour. The zinc plates must be firmly screwed to the covers, and must lie parallel with the side of the cells and hang perpendicularly. The cover may then be laid on the top of the cell, note being taken that the indiarubber packing is adjusted in position carefully, and that the zinc does not touch the sides of the cell. The cover screws should then be tightened carefully, and, in doing so, it is desirable that they are screwed down turn and turn about, so as to obtain an even pressure on the indiarubber packing.

When, in making up a battery of Neotherm cells, the individual cells have to be placed close together, insulating material, such as asbestos sheet, should be placed between each two cells, as, if the iron bodies are not insulated from each other, the cells will be short-circuited when they are connected in series. Connecting wires should be of ample gauge, so that a drop of potential is avoided as far as possible. Before wires are attached to the centre terminals, it should be ascertained that the ebonite bushes do not turn in the covers. If the cells are required to be transported while they are filled with solution, the vent tubes on each should be covered with the indiarubber sleeves provided for the purpose, thus preventing the escape of the liquid. These sleeves must be removed before the cell is taken into use, and, as a precaution, it is desirable that a thin wire be passed down the vent tube to remove any soda which may have accumulated there during transit. The open circuit E.M.F. of these cells is nominally 1·1 volt; but should this be found to be higher, no importance need be attached to the fact, as the voltage will fall as soon as the battery is giving a current; as soon as the voltage has fallen to 0·4 per cell, the battery requires regenerating, but no harm will be done to it if the voltage is allowed to fall below this figure.

For regenerating a Neotherm cell, loosen the cover and lift it, with the zinc attached. Clean the zinc by rinsing it in clean water, and remove any zinc oxide. A wire brush is a convenient tool for this purpose. The zinc plates should then be dried and set aside until required. The solution should not be removed from the cells until the oven or stove in which they have to be heated has reached the necessary temperature. When the electrolyte is emptied from the cells all the residue must be washed out with it. If there is not sufficient solution wherewith to rinse the cell thoroughly, other solution must be used, perhaps from another cell, but on no account must water or other fluid be employed for the purpose. The cells may now be placed in the oven with the open tops uppermost. Any form of oven in which the necessary temperature can be obtained will answer the purpose, except that it must not be one where the heat is generated by gas inside the oven itself. The time required for regeneration depends upon the temperature of the oven. At a temperature of 250° to 300° F. this should be completed in from 3 to 4 hours for the No. 1 cell, and in from 4 to 6 hours for the No. 2 cell. It is desirable that the doors of the oven be kept closed during the baking. The lining of the cells which, after discharge, is of a red colour, will be changed to a uniform black colour when regeneration is completed. When the cells have been removed from the oven and have cooled down, they may be recharged with fresh solution in the same manner as new cells, or they may be laid aside until such occasion as they may be required again. The zinc plates may be used again and again until they become too thin. If, when the cells are again set up ready for use, it is found that the E.M.F. on open circuit is below 1·0 volt, it is an indication that the regenerative heating has not been carried out completely, in which case it will be necessary to dismantle the cell and repeat the baking process.

The Combat Battery

In this type of primary cell the container and cover are of porcelain, and the copper oxide is held in a perforated metal cylinder. The cylindrical zinc outside the metal cylinder is supported by porcelain insulators fixed to the metal cylinder. The exciting fluid is a solution of caustic soda, and oil is poured into the top of the cell to prevent creeping and evaporation. The cell, with a jar 8 in. high, has a capacity of 300 ampere hours. Practically all the remarks made in connection with the Edison-Lalande apply in this case also.

CHAPTER VII

SILVER-ZINC AND PLATINUM-ZINC CELLS

The De la Rue Battery

THE original form of the De la Rue battery utilises a fairly stout silver wire to form the negative element, around which is cast a cylinder of fused chloride of silver. A zinc rod forms the other element, and the pair are placed in a glass jar or other receptacle. The exciting solution is a saturated solution of common salt (sodium chloride). The chemical action is briefly thus :—The zinc and salt react and yield sodium and a zinc salt containing two molecules of chlorine. The sodium thus produced, in the presence of water, immediately oxidises and becomes soda, liberating hydrogen. Hydrogen in its turn combines with the chloride of silver, with the result that metallic silver and hydrochloric acid are produced. The latter then combines with the previously liberated soda, and the result is ordinary salt and water. Thus the chemical energy is at the expense of zinc, and the silver salt is used up to thicken gradually the silver element. The cell gives 1·3 volts when in best condition, but owing to high internal resistance the current output is limited unless the cells are cumbersome and expensive. The cells are sometimes used for working small medical coils, but seldom where larger currents are required such as for lighting purposes. These cells were used in 1868 by Warren de la Rue and Müller to ascertain the nature of electric discharges in rarefied gases. No less than 14,400 small cells were used, coupled in series, to determine the striking distance of the spark in air.

Gaiffe's cell is very similar to the foregoing, except that copper is substituted for the silver element.

The Smee Battery

This was invented in 1840, and is still in use for electroplating, etc. A platinised silver plate (silver coated with platinum black) *Ag* (Fig. 57), is clamped between, and insulated from, two heavy zinc plates **Zn**, which have been amalgamated. The electrolyte is dilute

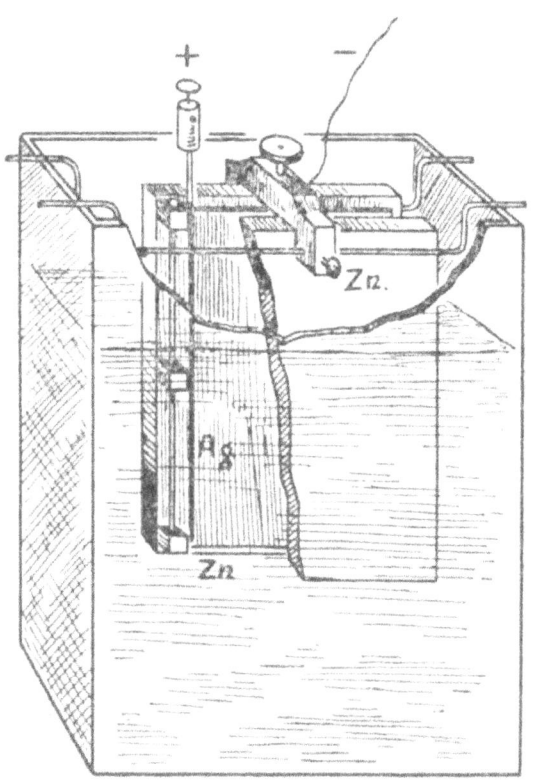

Fig 57.—Smee Cell.

sulphuric acid. The containing jar is much deeper than the plates so that the zinc sulphate formed may fall to the bottom and be clear of them. It is usual to connect up a number of these cells in series and provide an arrangement for lifting the plates out of the solution when the battery is not in use. The cells do not polarise so quickly as a zinc-copper cell, but the E.M.F. is only 0·5 volt.

The Grove Battery

The cell introduced in 1839 by Sir William Grove has been modified a number of times, but the present-day pattern is very much on the lines of the original one. The container is a glass jar in which is a porous pot surrounded by a stout, split, amalgamated

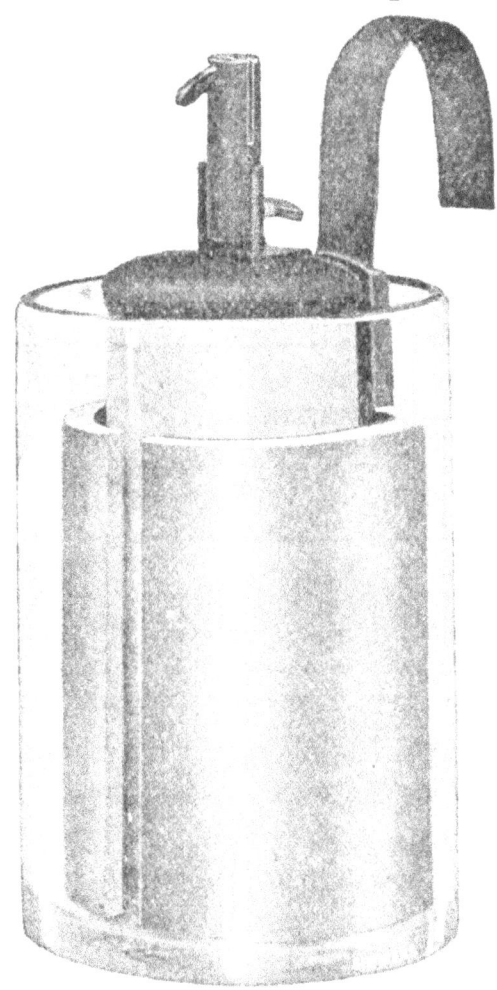

Fig. 58.—Grove Cell.

zinc cylinder terminating in a copper connecting strap. The porous pot contains a strip of platinum bent to S-section, exactly as in the original pattern; this electrode is attached to a cover which rests upon the porous pot. For charging the porous pot of the pint-size cell about $5\frac{1}{2}$ oz. of nitric acid (density 1·38) is necessary, and for the quart-size,

16½ oz. The outer cell, in which the zinc cylinder stands, is charged with dilute sulphuric acid to within ½ in. of the top, the level of the nitric acid in the porous pot being slightly higher.

The action of the battery is as follows :—The action between the dilute sulphuric acid and the zinc causes the formation of zinc sulphate and the decomposition of the water; the hydrogen thus left free passes through the porous pot and takes up some of the oxygen of the nitric acid, with the result that water and the objectionable fumes of nitric oxide are formed.

The pattern of Grove cell above described is made in the following sizes :—

Size.	E.M.F. approx.	Internal Resistance, approx.	Weight without Charge, approx.	Overall Dimensions, approx.
	volts.	ohms.	lb.	in.
1-Pint	.. 1·9	0·20	2¾	7 high × 3¾ dia.
2-Pint	.. 1·9	1·15	6¼	9 high × 4½ dia.

CHAPTER VIII

"DRY" CELLS

THE dry cell is dry only in the sense that it does not spill. The chemical action, and the chemicals used, do not differ materially from those of the Leclanché battery. Dry cells in practical use are nearly all some form of sal-ammoniac cell with the zinc made into the form of a containing jar. The top is sealed over to prevent evaporation, and the electrolyte is all taken up by an absorbent, so that there is no liquid to spill even before the top is sealed over. The carbon is usually in the form of a slab.

The Gassner Cell

This was one of the earliest dry cells to be made with a thin sheet zinc case, and though it is now practically obsolete, brief particulars of it will be useful. It was made in practically any size and shape—oblong, circular, or oval. The zinc case was the positive element, and was nearly filled with zinc oxide and gypsum, made into a paste with a zinc chloride solution. The negative element in the centre was a capped carbon bearing a binding screw. The contents of the cell were sealed over with a composition resembling marine glue. In one form of this cell the case was of square form and divided by a zinc partition into two equal parts in each of which was a carbon block. The object was to ensure a low internal resistance. In a small cell of the Gassner type, a hollow carbon cube was used; in another form the zinc case was enclosed in an enamelled iron vessel.

The Lessing Cell

Dr. A. Lessing's cell is shown in section by Fig. 59. It has an outer vessel P of porcelain, immediately inside which is a sheet zinc cylinder z in one piece, with which is the strip T, which connects to the negative binding screw w ; thus soldering or riveting inside the cell and consequent risk of local action are avoided. The negative plate c is a flat carbon surrounded by manganese dioxide contained in a canvas bag B bound round with thread. The electrolyte—sal-ammoniac solution thickened with flour, etc.—occupies the space between the bag and

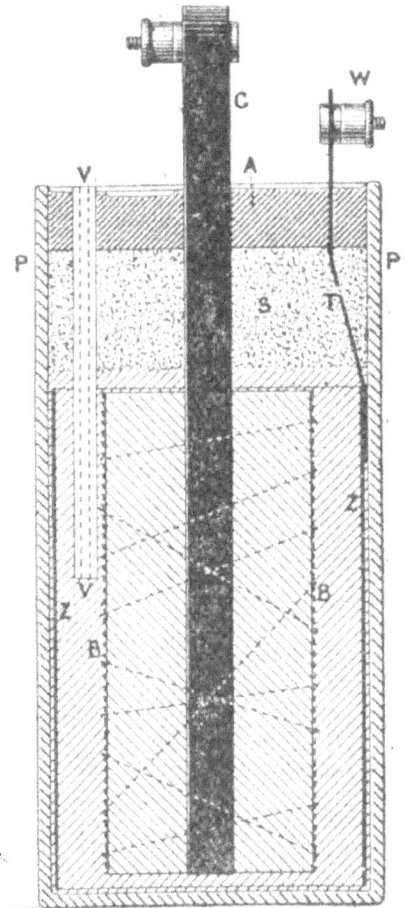

Fig. 59.—Section of Lessing Cell.

the zinc. There is a layer of sawdust s, and above this a bituminous seal A. The vent tube v is of lead. This is one of the historical cells which, although it gave very promising results, has now been largely superseded.

The E.C.C. or Burnley Cell

This has been in general use for many years, and in its improved form is an economical and efficient device. It has long given good service on bell and telephone systems. The construction of the older type is shown in Fig. 60, in which z is the outer case insulated at the bottom by a millboard case I. In the modern form of the cell, the millboard covers the whole of the zinc case. The carbon c has a brass terminal T, and is surrounded with

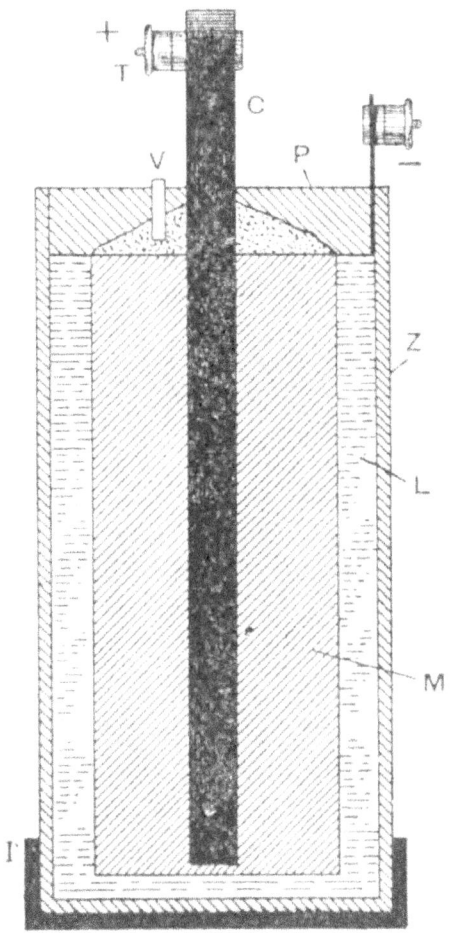

Fig. 60.—Section of E.C.C. or Burnley Cell.

a paste M of manganese peroxide and powdered carbon moistened with a solution of sal-ammoniac and zinc chloride. This is, in turn, surrounded with a paste L of plaster-of-Paris and flour moistened as before. The whole is sealed with bitumen P, through which passes the vent tube v.

The improved form of the cell is made up with millboard insulating cylinders which are waxed inside; a special method of sealing is employed, and the capacity is claimed to have been very considerably increased. It is obtainable in the following sizes from the Electrical Power Storage Company:—

Size.	*Overall Dimensions.*	*Weight.*	*Ampere Hour Capacity.*
	in.	lb. oz.	
Chocolate Label:			
No. 1 Round	8 by 3¼ diam.	3 13	100 amp. hrs.
,, 2 ,,	7 by 2½ ,,	1 14	48 ,, ,,
,, 3 ,,	5¾ by 2⅛ ,,	1 4	28 ,, ,,
No. 00 Square	8 × 4¼ × 4¼	9 5	180 amp. hrs.
,, 0 ,,	6 × 4¾ × 2½	3 13	78 ,, ,,
,, 1 ,,	8 × 3¼ × 3¼	4 3	110 ,, ,,
,, 2 ,,	7 × 2½ × 2½	2 3	60 ,, ,,
,, 3 ,,	5¾ × 2⅛ × 2⅛	1 9	34 ,, ,,
,, 4 ,,	4¼ × 1½ × 1½	0 9½	15 ,, ,,
,, 6 ,,	2¾ × 1⅞ × 1	0 5	6 ,, ,,
Orange Label:			
Extra-Sec A.	6¾ × 2⅝ × 2⅝	2 11	42 amp. hrs.
,, ,, B.	6½ × 2½ × 2½	2 8	38 ,, ,,
,, ,, C.	5¾ × 2⅛ × 2⅛	1 9	32 ,, ,,
,, ,, D.	6½ × 1¾ × 3	2 5	38 ,, ,,

Obach Cell

Dr. Obach's cell is of the Leclanché type wherein the exciting element, in solution, is mixed with inert material to form a stiff paste and fulfills the same function as the exciting solution in the ordinary fluid cells. The positive, or zinc, element is in the form of a pot and acts as a container for the other elements. The negative elements consist of a carbon rod surrounded by a depolariser composed of a mixture of peroxide of manganese and carbon, both in a pulverised condition. The exciting element consists, as in the Leclanché cell, of sal-ammoniac which, in solution, is mixed with flour and plaster-of-Paris to form a paste. The carbon rod, with its depolariser, is placed in the centre of the zinc box, and the paste is packed into the space between them. A space left at the top of the depolariser is filled with sawdust or similar material, and the top of the battery is covered by a layer of bitumen. The space at the top of the depolariser forms a receptacle for gases which are generated during the working of the cell, and it is ventilated by means of two small glass tubes in the cover. Terminals are attached to the zinc and carbon, and usually the cell is placed in a cardboard box. A terminal is attached to the carbon element, and either a terminal or length of connection wire to the zinc element according to requirements.

Obach cells are made in a variety of sizes and shapes, some being especially intended for motorist's use, but the following table gives particulars of

Size.	*E.M.F. approx.*	*Internal Resistance, approx.*	*Weight, approx.*		*Overall Dimensions, including Terminal, approx.*
	volts.	ohms.	lb.	oz.	in.
Round	*Cells :*				
J. ..	1·5	0·10	9	1	$4\frac{3}{8}$ dia. × $9\frac{5}{8}$ high
A. ..	1·5	0·15	4	0	$3\frac{5}{16}$ dia. × 8 ,,
B. ..	1·5	0·20	2	6	$2\frac{11}{16}$ dia. × $7\frac{1}{8}$,,
C. ..	1·5	0·25	1	6	$2\frac{1}{4}$ dia. × 6 ,,
D. ..	1·5	0·30	0	14	2 dia. × 5 ,,
Square	*Cells :*				
M. ..	1·5	0·15	9	10	$4\frac{7}{16}$ × $4\frac{7}{16}$ × $8\frac{1}{4}$
N. ..	1·5	0·15	5	0	$3\frac{3}{8}$ × $3\frac{3}{8}$ × $7\frac{1}{2}$
O. ..	1·5	0·20	3	0	$2\frac{13}{16}$ × $2\frac{13}{16}$ × $6\frac{5}{8}$
P. ..	1·5	0·25	1	14	$2\frac{1}{4}$ × $2\frac{1}{4}$ × $6\frac{1}{8}$
Q. ..	1·5	0·25	1	6	2 × 2 × $5\frac{1}{2}$
R. ..	1·5	0·30	0	14	$1\frac{13}{16}$ × $1\frac{13}{16}$ × 5
S. ..	1·5	0·50	0	10	$1\frac{1}{2}$ × $1\frac{1}{2}$ × $4\frac{1}{2}$
T. ..	1·5	0·50	0	6	$1\frac{1}{4}$ × $1\frac{1}{4}$ × $3\frac{5}{8}$
Oblong	*Cells :*				
U. ..	1·5	0·18	3	12	$3\frac{9}{16}$ × $2\frac{1}{2}$ × $7\frac{1}{2}$
W. ..	1·5	0·18	3	8	$2\frac{15}{16}$ × $2\frac{3}{8}$ × $7\frac{5}{8}$

those cells for general use which are stocked by Messrs. Siemens.

In the form of the Obach cell shown in Fig. 61, A is the outer vessel of zinc mounted on an insulating base B; C is a carbon rod surrounded by depolarising mixture D; between D and A is the electrolyte E of plaster, flour and sal-ammoniac; F is a paper ring, G a layer of ground cork, etc., H a second paper ring, K a bituminous seal, and L a glass tube to act as a gas vent. M is the terminal for the zinc; the terminal for the carbon is constituted by the screw P and nuts Q and R.

Dura Cell

In this cell the carbon, with its depolariser, is wrapped with a thin layer of textile material and bound and tied with string. The exciting layer constitutes the special feature of the cell. The materials of which this layer is composed, instead of being placed in the cell in a moist condition, are made up in the form of a specially prepared powder which is packed into the space between the depolariser and the zinc. To render the cell active it is only necessary to pour plain water through a hole at the top and, after a few hours, the exciting salt is dissolved, the liquid becomes completely absorbed, and the cell is then in the exact condition of an

Size.	E.M.F. approx.	Internal Resistance, approx.	Weight, approx.		Overall Dimensions, including Terminal, approx.
No.	volts.	ohms.	lb.	oz.	in.
53 ..	1·50	0·13	7	12	$4\frac{7}{16} \times 4\frac{7}{16} \times 8\frac{1}{8}$ high
54 ..	1·50	0·15	4	3	$3 \times 3\frac{3}{4} \times 7\frac{1}{2}$,,
55 ..	1·50	0·18	2	8	$2\frac{7}{8} \times 2\frac{13}{16} \times 6\frac{5}{8}$,,
56 ..	1·50	0·20	1	7	$2\frac{1}{4} \times 2\frac{1}{4} \times 6\frac{1}{4}$,,
57 ..	1·50	0·23	1	2	$2\frac{1}{16} \times 2\frac{1}{16} \times 5\frac{1}{2}$,,
59 ..	1·50	0·50	0	8	$1\frac{1}{2} \times 1\frac{1}{2} \times 4\frac{1}{2}$,,
62 ..	1·50	0·15	2	12	$2\frac{1}{16} \times 2\frac{7}{16} \times 7\frac{5}{8}$,,

ordinary dry cell when it is first made up. In cases where it is likely that cells may be kept in store for a number of years, this type possesses the advantage that, if left uncharged with water until such time as it is taken into use, it will remain absolutely inert and will suffer no loss. It is made by Messrs. Siemens in the sizes given in the table on p. 96.

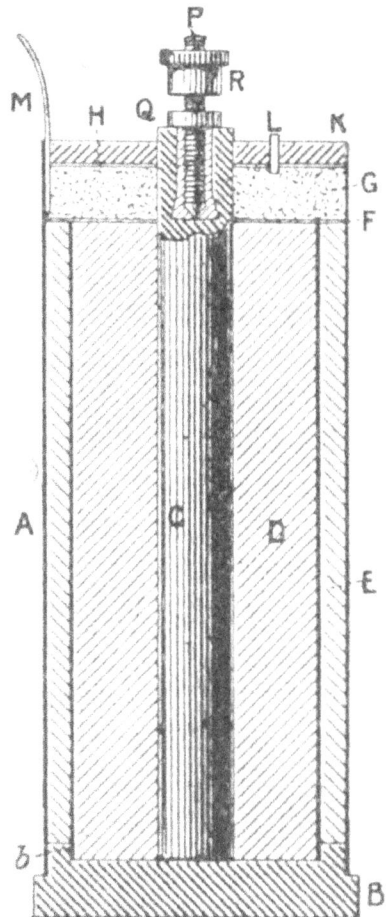

Fig. 61.—Section of One Form of Obach Cell.

Hellesen Cell

In this cell, as in the others, the zinc element is a cylinder having a connection wire or terminal attached thereto. The negative element consists of a carbon rod fitted with a terminal and surrounded by a depolariser composed of pulverised peroxide of manganese and carbon. The depolariser is wrapped around by a thin textile material and bound and tied with string. The exciting element consists of chloride of ammonium (sal-ammoniac) which is mixed in solution, with flour and plaster to form a paste or with material of a gummy nature to form a jelly. The negative element is placed in the centre of the zinc cylinder and the exciting paste filled into the space between them. A thick layer of plaster-of-Paris is then poured over the top. The battery is placed in a square cardboard box, the corners between the round zinc and cardboard box being filled with sawdust. The top of the cell is covered with a layer of bitumen. To provide a method of allowing the escape of gases which form during the working of this cell two or three small holes are made in the zinc, near to the top, and within the space covered by the layer of plaster-of-Paris. The gases percolate through the plaster and pass through the holes in the zinc into the sawdust. A long glass tube extends from the bitumen cover nearly to the bottom of the box. The gases pass down through the sawdust to escape by this tube, and any moisture that may be carried over with the gases remains in the sawdust.

The Hellesen dry cell is frequently made up in the following way :—It has a double zinc case, one within the other, both of them being perforated, the inner one near the top and the outer near the bottom. Slag wool or carefully selected sawdust free from all resinous matter is used to fill the space between the two cases, and acts as a kind of ventilating arrangement for the escape of gases. There is a central carbon rod in the inner zinc compartment with a muslin bag tied to the lower half. This bag is filled with the depolarising paste of black oxide of manganese, carbon or other substance that experience has shown to effect any improvement in working. The remaining space in the inner zinc compartment is filled with a jelly (either gelatine or agar-agar) containing the exciting salts, such as sal-ammoniac, chloride of calcium, or zinc chloride. Then comes a final layer of plaster-of-Paris reaching just over the perforations near the top, and a layer of pitch to seal the whole. A wire projects through the pitch from the inner zinc, and also a terminal from the carbon rod, to provide the necessary connections. Any gases given off escape through the plaster-of-Paris and slag wool between the two cases. The E.M.F. should be 1·5 volt, and internal resistance 0·2 ohm, in the standard-sized cell of 3 in. by 3 in. by 7½ in.

The Hellesen cell has been adopted for the ignition of automobile motors. The necessary combination includes ignition coil, positive make-and-break contact, sparking plug and a battery of dry cells. The ignition coil is of the plain non-trembler type, with primary and secondary windings. The design of the mechanical contact ensures a quick and reliable make-and-break action during a very brief period, a hot spark being produced with a low consumption of current. The batteries are made up in compact sizes, and it is claimed that they have low internal resistance, long life, and great capacity. It is recommended that two sets of cells be carried, one set being in use for about one hour whilst the second set is recuperating, the change-over being effected with a small two-way switch. It may be of interest to note here that the well-

Size.	E.M.F. approx.	Internal Resistance.	Weight, approx.		Overall Dimensions including Terminals.
No.	volts. about	ohms. about	lb.	oz.	in.
1 ..	1·5	0·20	5	12	4 × 4 × 7¾ high
2 ..	1·5	0·20	3	0	3 × 3 × 7¼ „
3 ..	1·5	0·25	1	14	2⁹⁄₁₆ × 2⁹⁄₁₆ × 6¼ „
4 ..	1·5	0·30	1	9	2¼ × 2¼ × 6¼ „
6 ..	1·5	0·50	0	8	1½ × 1½ × 4½ „

known De Dion-Bouton firm has for many years employed dry batteries in the place of accumulators on its small cars.

Messrs. Siemens, who have brought the Hellesen cell to a high state of efficiency, make it in the sizes given in the table on p. 99.

The Dewa Dry Cell

The Dewa cell was put on the market by a German firm. It contains a carbon rod which is hollow for most of its length, so that communication is made with the outer air by a special tube passing through the seal. The depolariser surrounding the carbon rod is ventilated through the hollow in the latter, and is also ventilated through a tube which projects into a hollow space above the depolariser. In one type of cell, vertical channels have been provided in the depolariser to allow of increased air supply and to promote the expulsion of the gas. In some cases a special ventilating space is also placed at the bottom of the cell. This system of ventilation, it is claimed, allows any free hydrogen to be quickly removed from the carbon, while at the same time the depolariser is provided with fresh supplies of oxygen, thus increasing both the efficiency and life of the cell.

Making Dry Cells at Home

Amateurs not infrequently ask questions which seem to imply that reliable dry cell making is a common everyday practice, and can be taken up by anyone without experience. Such an idea is very erroneous. When, after long and costly research, a good cell is produced, the knowledge of the process has a high market value, and an attempt is made to keep secret the precise details of the process. Thus it follows that the makes of good dry cells are few. The following, however, is based on experience, and may be adopted as a guide in the home-manufacture of dry cells.

Cells should be made of sheet zinc, and may be of any size or shape. The zincs form the positive elements of the battery, and the negative elements consist of strips or rods of carbon, which are fitted with brass caps. Each carbon strip or rod must be very tightly packed into a small canvas cell with a paste composed of finely powdered carbon and manganese dioxide, of each 5 parts; sal-ammoniac and chloride of zinc, of each 1 part; glycerine, ½ part; water, 1 part. This cell, with its carbon, must be put in the centre of a zinc cell, which must then be nearly filled with a paste composed of one teaspoonful of flour dissolved in a solution of sal-ammoniac, ½ oz.; zinc chloride, ¼ oz.; glycerine, ½ oz.; water, 4 oz. Rub the flour into a smooth paste first, then add the paste to the solution and boil the whole together, pouring it whilst hot into the zinc cells, and allowing to get quite cold. In the bottom of each cell a small piece of cardboard should be placed for the canvas cell to rest on, and the cells should be sealed with a hot mixture of pitch, resin, and paraffin wax in equal parts, poured from an iron ladle. When the pitch seals have cooled hard, the seal of each cell must be pierced with a hot iron wire, in order to form vent holes for the gases. A short piece of No. 24 copper wire must be soldered to each zinc cell. The zinc of No. 1 cell must be connected to the carbon of No. 2 cell by soldering the connecting wire to the brass cap of this cell, and the rest of the cells must be connected in a similar manner, thus leaving one carbon cap at one end of the battery, and one wire from the zinc at the other end of the battery free for connection to the apparatus that is to be worked. A battery of this kind will not deteriorate by remaining idle and will continue in working order until the charges in the cells are exhausted. The period of time during which the cells will give a full current will be governed by (a) the amount of the charge; (b) the size of each cell; and (*c*) frequency of employment. The E.M.F. of each cell will be that of a wet Leclanché cell, namely, 1·6.

Making a Flash-lamp Battery

A tiny flash-lamp costs about ninepence, and the case and battery can be made at a cost of a few pence. In the first place, obtain a piece of thin sheet zinc, 1 ft. long by 6 in. wide, and cut it into three pieces 6 in. by 4 in., and from each of the pieces cut off a 1 in. by 4 in. strip. With these pieces make three zinc cells—one 1 in. square, the others D-shape, as shown in plan at Fig. 62. Solder the edges, and fit a zinc bottom to each, cut from one of the narrow strips; then solder all joints watertight. Having made the cells, solder a 2-in. length

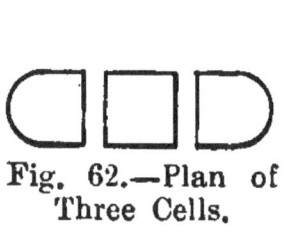

Fig. 62.—Plan of Three Cells.

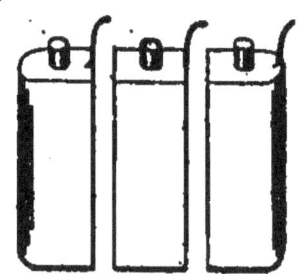

Fig. 63.—Three Cells with Terminals.

of electric bell line wire to each, as shown at Fig. 63. Next obtain three rods of carbon, of any section (electric arc lamp carbon will do), and cut three pieces to the same length as the zinc cells; then fit a small brass cap or collar tight to each carbon.

Next make three small canvas bags to hold the carbon and form the inner cells; the bags must be as long as the carbon rods, and just large enough to go into the zinc cells and leave a little space around them when full. The carbon rods must be packed tight in these bags with a pasty composition made according to the first recipe in the previous paragraph. Dissolve the sal-ammoniac and zinc chloride in the water, add the glycerine, then make the whole into a stiff paste, which must be rammed in tight around

each carbon rod. The powdered carbon in this paste
is merely that of crushed battery plates or of arc-
light rods. When the bags are full, tie each one
tight around the top, under the cap on the carbon,
with strong waxed twine. Put a bit of cardboard
in each zinc cell for the bags to rest on, then put
in the filled bags, and fix them in the centre of each
cell with the paste mentioned in the foregoing paragraph. Boil the mixture, and pour it into the cells
whilst hot; then set aside to get cold. When the
mixture has became quite cold and firm, solder the
wire from one zinc to the carbon rod of the next
cell, and the wire from the zinc of this to the carbon
rod of the next, thus forming two loops of wire,

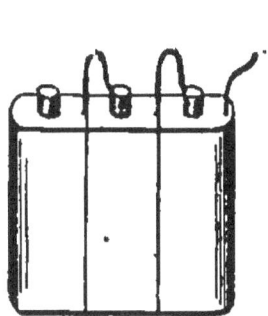

Fig. 64.—Cells connected
Together.

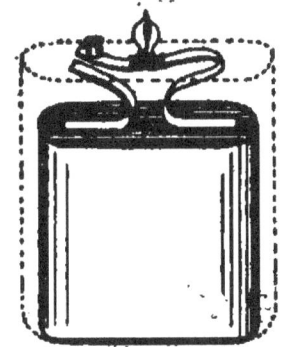

Fig. 65.—Flash-lamp Battery
Complete.

and leaving one end free, as in Fig. 64, which shows
the battery ready for sealing.

Now obtain two strips of paraffined cardboard
to separate the cells, and some strong black paper
to form a case for the battery. This case is made
by pasting several folds of the paper round the
three cells with the two strips of cardboard in
between them. The case should come ½ in. above
the rims of the cells. Now solder a V spring of hard
sheet brass or German silver to the free carbon
cap, and another to the zinc cell at the other end,
or to the wire attached to the zinc (*see* Fig. 65).
These springs will make contact with other strips
of brass connected to the lamp in the cover of the

lamp case. This done, seal the cells with a composition made of pitch, resin, and paraffin wax in equal parts, made hot in an iron ladle, and poured whilst hot. When this has set cold, pierce each cell with a red-hot wire to form vents for the gases generated in the cells when the battery is working. This battery is made to fit into a tin case having a small pea lamp screwed into a depression in the cover, to which is also attached springs for connecting the lamp to the battery. One of these springs connects only when it is pressed down by the push-button. An outline of the lamp, case and cover is shown by the dotted lines in Fig. 65. The stem of the lamp has a spiral groove on its surface, and this fits into a socket formed in the lid of the lamp case.

The light from a 4-volt pea lamp equals from $1\frac{1}{2}$ to 2 candles. It is advisable to use the lamp as a flash lamp only, just to see the time by a watch in the night, or to look in a dark cupboard for one or two minutes at a time. However, the battery recovers strength within a few minutes, even if the light has been kept on until it became dim. Each time the lamp is used, some of the oxygen in the battery is consumed, and eventually the stock becomes exhausted. The length of time it will last depends on the use made of the lamp. It is usually good for 1,000 flashes. The contents of these dry cells will part with their water and become dry and hard after being in use for some time. They will get in this condition more rapidly if the tops of the cells have not been sealed with the pitch mixture, if they are exposed to a high temperature, and after they have done much work (just before complete exhaustion).

The amalgamation of the thin zinc sheet used for dry battery cases is neither practicable nor advisable. It can only be done in the case of substantial battery plates $\frac{1}{8}$ in. thick and upwards, as the mercury quickly attacks and amalgamates with the zinc sometimes penetrating to a considerable depth; the zinc then goes rotten and crumbles. This fact, indeed, is sometimes taken advantage of in cutting up battery plates from the sheet; grooves are scored on both sides of the sheet where required to be cut, and mercury poured in them. In a few moments the mercury will have eaten into the zinc sufficiently to enable the piece to be broken off easily.

Life of Dry Cells

The life of a dry battery is governed by: (a) The quantity of oxygen-making material put into the cells; (b) the rate at which the store of oxygen is consumed. The carbon element is indestructible, and therefore never wears out. The zinc element outlasts the charge put in the cells. (a) The oxygen-making material in the charge of a dry battery, and also in any other form of Leclanché battery, is peroxide of manganese. Large cells have a large capacity for this material, and therefore contain a larger store of oxygen than small cells. (b) Peroxide of manganese is only slowly broken up by the electric current, so can only maintain a small volume of current. Therefore, if the battery cells are short-circuited through a conductor of low resistance, the current becomes faint for want of oxygen, and, if this short-circuiting is often repeated, the store of oxygen is soon exhausted. The same results follow a long-sustained action of the battery. Therefore, if a dry battery is often used on an alarm bell to ring it for long periods, it will have a short life, although otherwise it is one of the best batteries for the purpose.

There is only such a small chemical action in a dry cell when the circuit is open that such action may be neglected altogether. A dry cell may be said to be at rest when not in use, and then will remain in working order for an indefinite time, unless injured by exposure to heat.

Re-charging Dry Cells

Dry cells cannot be re-charged properly, but when, as often happens, they fail through lack of moisture they may sometimes be revivified by boring a small hole in their cases and standing them in a vessel of clean water for about 24 hours. But this is quite uncertain. A better method is to drill a $\frac{1}{2}$ in. hole through the top sealing, removing some of the composition to within 1 in. of the bottom of the cell, and then to pour in as much of a solution of sal-ammoniac as the cell will absorb. At best this only slightly lengthens the life of the cell.

The following detailed instructions on re-charging are of general application, although they refer particularly to cells of the Dania make :—Provided the zinc is still good, exciting paste and carbon mixtures are required, the quantities, of course, being regulated according to the size of the cell. The carbon mixture should be 2 parts of crushed carbon and 1 part of manganese made damp with zinc chloride or sal-ammoniac solution. This mixture must not be made too damp, or it will not pack well. The exciting paste is made of gum tragacanth, which should be made up in a small quantity at a time, say $\frac{1}{4}$ oz. ; it is put into an ordinary bowl, and with a round piece of wood or pot-stick, brought to the consistency of cream by the addition of a little glycerine. A solution of sal-ammoniac having been previously prepared to the strength for Leclanché cells, about $2\frac{1}{2}$ oz. to 1 pint of water, this is very gradually stirred into the tragacanth mixture. The mixing should continue as long as the gum shows any signs of thickening, till it becomes about as thick as table jelly, and then be set aside.

It is advisable to make fresh cylinders of zinc, soldered up one side and covered round with three or four layers of brown paper. Another zinc cylinder is made $\frac{1}{8}$ in. or $\frac{3}{16}$ in. less than the zinc which forms the cell, a good overlap being allowed. This may be formed by bending the zinc round any suitable roller, and tying it with wire to keep the ends together. Set this on a table or other flat surface, pour in a little of the prepared black mixture, and set in the carbon, which is preferably a round one. Having placed the carbon in position, hold the cylinder on the bench, and at the same time pack in the mixture as tightly as possible. When within about 1 in. of the top, remove the wire, and the zinc will spring open, leaving a nicely moulded cylinder of carbon and manganese. Next cut a piece of muslin or fine calico to cover this, overlapping about ½ in., with the same at top and bottom. This covering should be slightly damped, to make it fit better. Roll up neatly, and wind tightly with thin cord from top to bottom, thus pressing the mixture more closely to the carbon pencil in the centre. Little pegs of wood will keep this from touching the zinc, but may be dispensed with if the exciting paste has been made thick enough.

All that now remains to be done is to put a little of the exciting paste into the zinc pot and press down the carbon. Turn it round once or twice to spread the paste evenly, and clean out whatever comes to the top. This completes the cell, which is now ready for sealing up. A circle of paraffined paper is put in, then some damp sawdust, another layer of paper, and, finally the melted pitch, which may be pure or mixed with plaster-of-Paris. The wire to the zinc pot should be soldered on, preferably on the inside, before the cell is set up, so that it does not interfere with the case when pressed in. A small glass tube is inserted for ventilation.

It is possible to re-charge an exhausted dry cell in the same way that an accumulator is charged, but the effect is merely temporary. A current from two or more Bunsen cells, for instance, can be passed through the cell from carbon to zinc for an hour or so ; but this is more or less a waste of current.

CHAPTER IX

STANDARD CELLS

THE Board of Trade defines the standard of electrical pressure in these words :—"The volt, which has the value 10^8 in terms of the centimetre, the gramme and the second of time, is the electrical pressure that, if steadily applied to a conductor whose resistance is 1 ohm, will produce a current of 1 ampere, and which is represented by $0{\cdot}6974 \left(\dfrac{1000}{1434}\right)$ of the electrical pressure at a temperature of 15° C. between the poles of the voltaic cell known as Clark's cell, set up in accordance with the specification appended hereto."

In the words of the specification, "the cell consists of zinc or an amalgam of zinc with mercury in a neutral saturated solution of zinc sulphate and mercurous sulphate in water, prepared with mercurous sulphate in excess."

To obtain standard results, all the materials used must be chemically pure ; the zinc and the mercury are procured by redistillation of the purest metal obtainable.

In preparing the mercurous sulphate, mix the pure chemical as purchased with a small quantity of pure mercury, and wash the whole in distilled water two or three times, draining off the water thoroughly. The zinc sulphate solution is obtained by treating neutral saturated solution of pure recrystallised zinc sulphate with distilled water with twice the weight of crystals, and then adding zinc oxide (about 2 per cent. by weight) to neutralise any free acid. Dissolve with a gentle heat not exceeding 30° C. Add 12 per cent. of the mercurous sulphate previously prepared, which will neutralise any free zinc oxide, and lastly filter the solution while still warm into a clean receptacle.

The final preparation of the mercurous and zinc sulphate paste is accomplished by adding the two previous preparations together with some crystals of pure zinc sulphate and a small quantity of pure mercury. These are shaken up well together to form a cream-like consistency, and heated gently to a

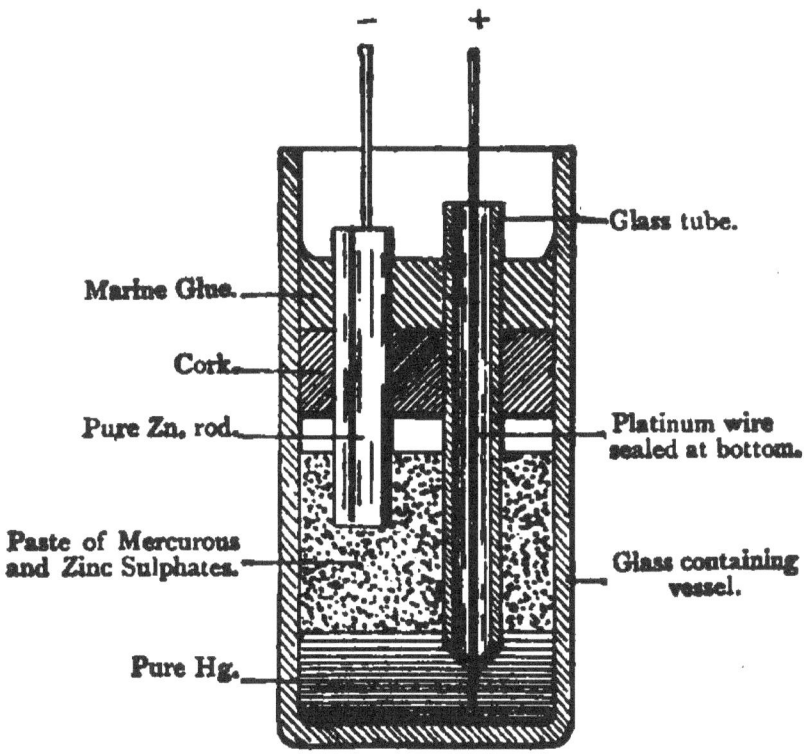

Fig. 66.—Board of Trade Standard Cell (Clark's).

temperature of 30° C. as before for an hour, keeping well stirred.

The method of setting up the cell is as follows (*see* Fig. 66) :—Take a test tube, 2 cm. diameter by 5 cm. deep, and place ½ cm. depth of mercury at the bottom. Cut a cork ½ cm. thick to fit the tube, and at one side bore a small hole, through which the zinc rod can pass tightly. At the other side is another hole for a glass tube, which covers the platinum wire. A small nick at the edge of the cork will allow air to pass when it is pushed down the tube, and the cork requires soaking in hot water to thoroughly cleanse and soften it before use.

The platinum wire making contact with the mercury need not be larger than No. 22 gauge, and is sealed into one end of the glass tube to prevent it making contact with the other materials, the free end passing up the glass tube to form one terminal. Clean the glass tube and platinum wire carefully, heat the exposed end of the platinum to a red heat, and insert it well under the surface of the mercury. Then shake up the paste, and introduce about 1 cm. depth on top of the mercury without soiling the sides of the test tube. Lastly, push in the cork with the zinc rod projecting downwards about 1 cm. The cork must go down until in contact with the liquid, expelling all air. The cell should be left for 24 hours before sealing up.

To do this, melt some marine glue until it is fluid enough to pour into the test tube to a depth of 1 cm. or 2 cm. If a more permanent sealing is required, the marine glue may be surmounted with a further cap of sodium silicate, and left to harden.

In using this standard cell, it should be so arranged that the cell can be immersed to the level of the cork in a water bath, and the exact temperature can then be ascertained with exactitude. The E.M.F. at a temperature of 15° C. is 1·434 volts, and this will vary with temperature, as in the following table :—

Temp.	Volts.	Temp.	Volts.
0° C.	1·450	15° C.	1·434
5° C.	1·445	20° C.	1·429
10° C.	1·439	30° C.	1·418

In the Fisher modification of the Clark cell, the mass of mercury, which is toublesome in the official pattern if the cell is roughly handled, is replaced by a small quantity of mercury contained in a cylindrical spiral continuation of the platinum wire (*see* Fig. 67). This is surrounded by a paste of mercurous sulphate, on the top of which are crystals of zinc sulphate and a saturated solution of zinc sulphate. The zinc rod passes down through

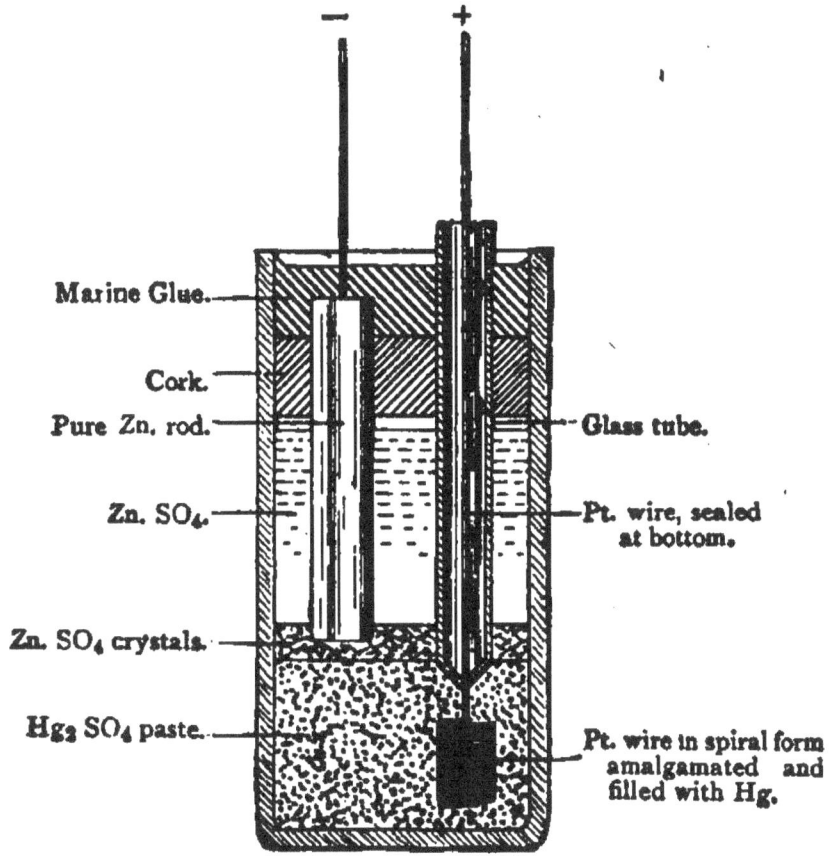

Marine Glue.

Cork.

Pure Zn. rod.

Glass tube.

Zn. SO₄.

Pt. wire, sealed at bottom.

Zn. SO₄ crystals.

Hg₂ SO₄ paste.

Pt. wire in spiral form amalgamated and filled with Hg.

Fig. 67.—Fisher's Modification of Clark's Cell.

the solution and terminates in the zinc sulphate crystals. It is claimed that cells set up according to this pattern give more concordant results than the official cells and that they stand rough usage, both electrical and mechanical, short of actual breakage, very well.

A great disadvantage of the Clark standard cell is its high temperature co-efficient. More than one experimenter has improved upon it in this

particular by substituting cadmium for the zinc
and cadmium sulphate for the zinc sulphate. In
Dr. Henderson's form, shown in Fig. 68, the mercury
in the bottom of the test tube has the usual paste
of mercurous and cadmium sulphates in contact
with it; on this there rests some moist cadmium
sulphate crystals, above which is an amalgam of
cadmium, consisting of 1 part by weight of cadmium
to 6 of mercury. The connections to both mercury
and cadmium are made by fine platinum wires sealed
through narrow glass tubes in the ordinary way,
the expense of long platinum wires being obviated
by soldering on copper leads at the point *s* low down
in the tubes. The cell is sealed up with marine
glue. The E.M.F. of cadmium cells, when made
with pure chemicals, was found to be 1·0188 volt
between 10° and 20° C., and the temperature
coefficient 0·003 per cent. per degree. The cells
were found to have an appreciable time-lag when
subjected to large and sudden changes of temperature,
but such changes can easily be avoided in actual
practice. The effect of ordinary impurities in the
materials was found to be small.

Lord Rayleigh devised the H-pattern of standard
cell and this, for a long time, was regarded as the
most reliable. In an improved cell of this type,
improved standard voltaic cell, patented by R. O.
Heinrich, of Berlin, a diaphragm between the liquid
electrode and the electrolyte liberates any air that
may accumulate on the surface of the diaphragms
in position in the cells; it also enables the diaphragms
to be maintained permanently in position between
the electrode and the electrolyte. The diaphragm
is in the form of a double piston, having two perforated discs on a hollow vertical tube, the usual
porous packing being between the discs, the lower
of which is funnel-shaped to facilitate the escape
of the air through the hollow tube on entering the
diaphragm in place. In making the cell, the liquid

electrode is, as usual, poured first into the tube,
the space between the discs previously mentioned
is packed with a porous material, and the piston-
like arrangement is then inserted in the tube until
the funnel-shaped lower disc rests just above the
liquid electrode. The electrolyte is poured into

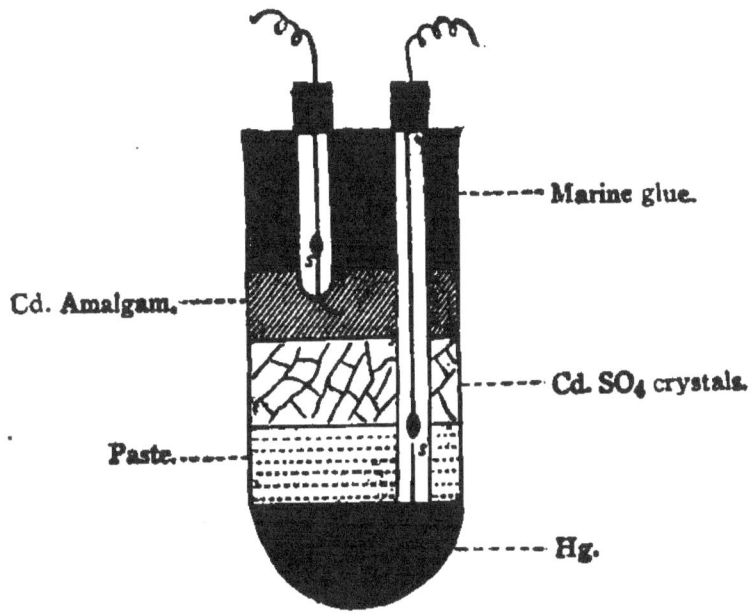

Fig. 68.—Cadmium Standard Cell.

the tube to nearly as high as the top of the hollow
central or vent tube, and corks are then inserted
and the cell sealed with paraffin wax. The cork rigidly
secures the vent tube and prevents any upward
movement of the diaphragms, thus maintaining them
in fixed relation with the liquid electrode.

H,

CHAPTER X

METHODS OF CONNECTING CELLS

"Series" and "Parallel"

CELLS can be arranged in an electrical circuit in either of two methods or in combination of these methods. If the cells are joined up so that the positive terminal of one is connected mechanically and electrically with the negative terminal of the next cell, and so on, as in Fig. 69, the end terminals being joined to the outer circuit, then the cells are said to be in *series*.

When, however, all the positive terminals of the set of cells are connected to a common bar or junction

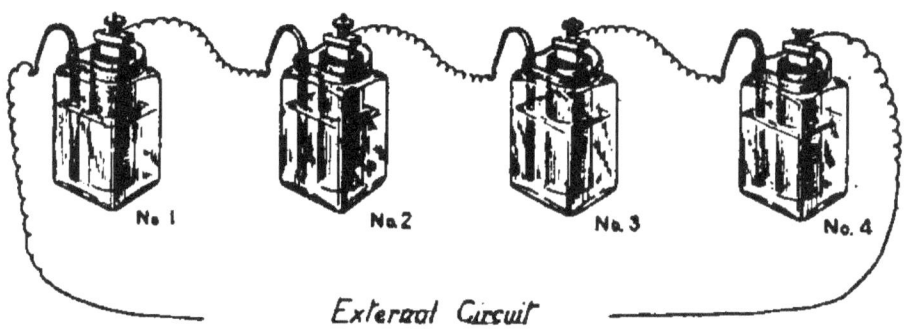

Fig. 69.—Cells Connected in Series.

and all the negative terminals to another bar, as in Fig. 70, the cells are said to be in *parallel*, and one end of the outer circuit would be brought to one bar and the other to the other bar or junction.

In the case of cells connected in series, the voltage or difference of potential between the outer terminals will be equal to all the separate voltages of the cells added together, so that, if the cells are all alike, it will equal the voltage of any particular cell multiplied by the number of cells in series,

In the parallel arrangement the voltage is that of one cell only, and for satisfactory working the cells should be of equal E.M.F.; the adoption of this system generally denotes that a large current is needed.

In the series system, if one cell is wrongly connected, as, for example, cell No. 2 in Fig. 71, its

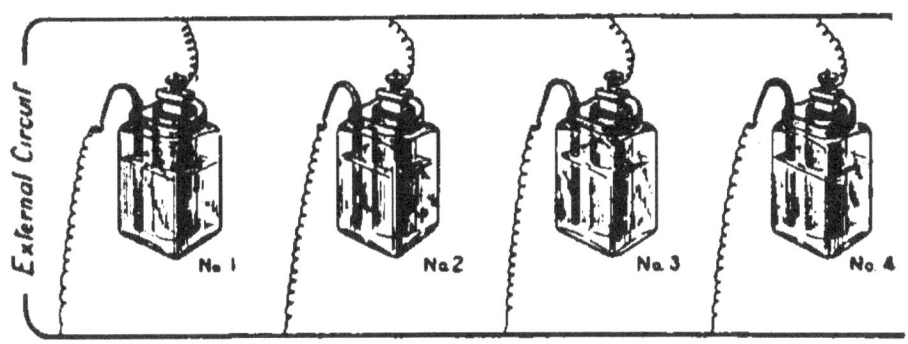

Fig. 70.—Cells Connected in Parallel.

E.M.F. will oppose that of the remainder of the cells, from which it must be subtracted in calculating the effective electro-motive force on the circuit. Thus, taking the E.M.F. of each cell as 1.5, the total E.M.F. of cells Nos. 1, 3, and 4 will be $1.5 \times 3 = 4.5$; but from this must be subtracted that of cell No. 2, 4·5 —

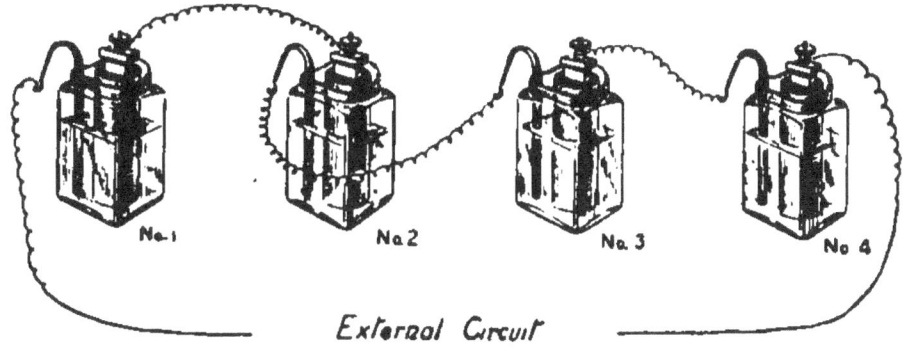

Fig. 71.—Cells Wrongly Connected.

$1.5 = 3$. The effect of the error in this instance is to halve the total E.M.F. obtainable.

E.M.F. and Potential Difference

The immediate object of using a cell or any electrical generator is to send a current through an external circuit. Now each cell has an appreciable internal resistance of its own, and a portion of the electrical pressure set up is employed on this resistance in " overcoming it," as it is expressed. A distinction is therefore sometimes made between the E.M.F. and the potential difference of a cell. The former is the total electrical pressure set up by the cell, and depends entirely on the materials used, whereas the potential difference of a cell is more properly regarded as the pressure that can be exerted on the external circuit, and equals the difference between the E.M.F. and the volts lost in sending the current through the cell.

As the volts lost in sending the current through the cell equals the internal resistance in ohms multiplied by the current in ampères, and as the E.M.F. is a fixed quantity, it follows that the potential difference varies with the current being produced, being greatest when no current is passing, and getting lower in value as the current is increased, until it is at a minimum when the cell is short-circuited.

The Volt

The volt is the unit of what may be called, for want of better terms, electrical pressure, potential difference, and electro-motive force, the last being usually shortened to E.M.F., the course adopted throughout this book. As stated already, the volt has been defined by an Order in Council, August 23rd, 1894, as " the electrical pressure that, if steadily applied to a conductor whose resistance is 1 ohm, will produce a current of 1 ampere," and also as " being represented by $\cdot 6974 \left(\dfrac{1,000}{1,434} \right)$ of the electrical pressure at a temperature of 15° C. between the poles of a voltaic cell known as Clark's cell " (*see* p. 108).

Typical Problems and their Solution

One or two problems may now be inquired into and their solutions worked out.

Suppose it is intended to set up one hundred Grove cells in series, but by mistake ten cells are arranged in opposition to the rest. What is the relation of the potential difference of the terminals on open circuit to that which would have been obtained if the mistake had not been made?

Here ninety cells are arranged to send the current in the required direction and ten cells to send it in the opposite direction; the net useful effect is therefore that of $90 - 10 = 80$ cells. The potential difference obtained will therefore be $\dfrac{80}{100}$, or ·8 of that which would result if all the cells were correctly connected.

The platinum and copper plates of a Grove and a Daniell cell are connected by a wire. Would there be a current if the zinc plates were also connected, and, if so, in which direction would it flow?

The arrangement is that of two cells in opposition, and the effective voltage would be the difference between the E.M.F.'s of the cells. For the Daniell cell this is from 1·07 volts to 1·14 volts, and for the Grove cell 1·9 volts to 1·95 volts. The effective E.M.F. would therefore be between ·8 volt and ·9 volt, the value of the current depending on the resistances of the two cells connected in series.

Six similar cells are arranged in series, and the circuit is completed through a length of wire and a galvanometer. The resistances of the battery, coil, and galvanometer are 10 ohms, 50 ohms and 20 ohms respectively. If the difference of potential between the terminals of the galvanometer is 2 volts, what is the E.M.F. of each cell of the battery?

Since the voltage necessary to send a given current through a given resistance equals that current in amperes multiplied by the resistance in ohms, and as in a series circuit the current is constant in value in all parts of the circuit, it follows that the voltages required vary directly with the resistances of the various parts of the circuit. Thus, if for a resistance of 20 ohms the pressure necessary is 2 volts, for a resistance of 10 ohms it will be $\frac{10}{20} \times 2 = 1$ volt, and for a resistance of 50 ohms, $\frac{50}{20} \times 2 = 5$ volts. The total pressure necessary is therefore $2 + 1 + 5 = 8$ volts, and, if the cells are all alike, the E.M.F. of each should be $\frac{8}{6} = 1\cdot33$ volts.

The terminals of a battery of five Grove cells, the total E.M.F. of which is 9 volts, are connected by three wires, the resistance of which is 9 ohms. The current through each wire is six-sevenths of an ampere. What is the internal resistance of each cell ?

The resistance of the external circuit—that is, of the three wires in parallel, is $\frac{9}{3} = 3$ ohms, and the total current is $3 \times \frac{6}{7} = \frac{18}{7} = 2\frac{4}{7}$ amperes. The pressure necessary on the external circuit is therefore $3 \times 2\frac{4}{7} = 7\frac{5}{7}$ volts ; leaving $9 - 7\frac{5}{7} = 1\frac{2}{7}$ volts for overcoming the internal resistance. And as the total current here is also $2\frac{4}{7}$ amperes, the internal resistance of the five cells in series must be $\frac{1\frac{2}{7}}{2\frac{4}{7}} = \frac{1}{2}$ ohm, so that the resistance of each cell is $\frac{\frac{1}{2}}{5} = \cdot1$ ohm.

One hundred battery cells, each having an E.M.F. of 1 volt and an internal resistance of 5 ohms, are to be joined up so as to send the maximum current through an external resistance of 125 ohms. How should this be done ?

For maximum current, the external resistance should equal the combined internal resistance. Now if there are x rows of cells in parallel, there will be $\frac{100}{x}$ cells in each row ; and the internal resistance of each cell being 5 ohms, the internal resistance of each row of cells will be $\dfrac{100}{x} \times 5$ in ohms, and of x such rows in parallel, $\dfrac{100}{x} \times \dfrac{5}{x} = \dfrac{500}{x \times x}$. But by the rule just given, this should equal 125 ohms, or $\dfrac{500}{x \times x} = 125$, whence $x \times x = \dfrac{500}{125} = 4$ and $x = 2$. Therefore two rows in parallel should be employed, each of $\dfrac{100}{2} = 50$ cells in series.

The Ampere

More useful exercises on cells can be attempted when the ampere, the unit of current or current-strength, has been considered. This has been officially defined as being " represented by the unvarying electric current which, when passed through a solution of nitrate of silver in water," in accordance with a given specification for a silver voltameter, " deposits silver at the rate of ·001118 of a gramme per second." According to the specification, the voltameter mentioned " measures the total electrical quantity which has passed during the time of the experiment, and, by noting this time, the time average of the current (or, if the current has been kept constant, the current itself) can be deduced."

Further, it is directed that for measuring currents of about 1 ampere, " the kathode on which the silver is to be deposited is to take the form of a platinum bowl not less than 10 centimetres in diameter and from 4 to 5 centimetres in depth." Also, the anode should be a pure silver plate about 30 square centimetres in area and 2 or 3 millimetres thick; it should be supported horizontally in the liquid, near the top, by a platinum wire passed through holes in opposite corners, and the anode should be enclosed in pure filter paper fastened at the back by sealing wax, to prevent disintegrated silver formed on the anode falling on the kathode. According to the specification, "the liquid should consist of a neutral solution of pure silver nitrate, containing about 15 parts by weight of the nitrate to 85 parts of water," and, to prevent changes in the resistance of the voltameter having too great an effect on the current, a resistance should be inserted, the total metallic resistance of the circuit being not less than 10 ohms.

How the Connections Affect Current

The laws as to current need not be entered into fully, but it may be stated that with cells connected in *series* the current in the outer circuit is that which could be supplied by one cell only (though not necessarily when acting on the same external resistance), and that in the *parallel* arrangement of cells the current in the outer circuit is the sum of the separate currents from the cells, the voltage being that of one cell only, as before stated.

It may also be added that in a series arrangement the current is the same in all parts of the circuit, and that in a circuit consisting of single conductors joined to portions consisting of several circuits the sum of the currents in the latter portions must equal the current in the single conductor parts. With the aid of these rules and of Ohm's law (*see* pages 14 and 15) several examples may be worked.

Further Problems Worked Out

A battery consists of twelve similar cells connected in series ; each has an E.M.F. of 1·1 (volts) and an internal resistance of 3 (ohms). If the poles of the battery are connected by a wire whose resistance is 240 ohms, what will be the strength of the current ? What will be the effect on the strength of the current of removing from the battery three of the cells, and replacing them with their poles inverted ?

In the first case, the total E.M.F. is $12 \times 1·1 = 13·2$ volts, and the total internal resistance of the cells in series is $12 \times 3 = 36$ ohms, therefore the total resistance of the circuit is $36 + 240 = 276$ ohms, and the current will be $\dfrac{13 \cdot 2}{276} = \cdot 05$ ampere (nearly). In the second case, the total resistance is of the same value as in the first case, but the voltage is altered, being that of nine cells in one direction and three cells in the opposite direction. It is therefore that of six cells, or $6 \times 1 \cdot 1 = 6 \cdot 6$ volts. The current will therefore be $\dfrac{6 \cdot 6}{276} = \cdot 042$ ampere (nearly).

A Daniell cell, the internal resistance of which is ·3 ohm, works through an external resistance of 1 ohm. What must be the internal resistance of another Daniell cell so that when it is joined up in series with the first and working through the same external resistance the current shall be the same as before? If the cells be joined up in parallel, how will the current be modified?

The total resistance with a single cell in circuit is $1 + \cdot 3 = 1 \cdot 3$ ohms, and by adding another similar cell the E.M.F. is doubled; therefore the total resistance must be doubled if the current is to remain constant; that is, the resistance of the second cell must be $1 \cdot 3$ ohms. Then, with the two cells in parallel, the internal resistance of the combination will be $\dfrac{\cdot 3 \times 1 \cdot 3}{\cdot 3 + 1 \cdot 3} = \dfrac{\cdot 39}{1 \cdot 6} = \cdot 24$ ohm, and the total resistance will be $1 \cdot 24$ ohms. Also, as the voltage is that of one cell only, the current with two cells in parallel will be to the current from one cell only as $\cdot 3 : \cdot 24$, or $\dfrac{\cdot 3}{\cdot 24} = 1 \cdot 25$ times the value.

CHAPTER XI

PRIMARY BATTERIES FOR ELECTRIC LIGHTING

Batteries too Costly for High Candle-power

THE average amateur is frequently under a misapprehension with regard to the possibilities of primary batteries for lighting purposes. We are led to believe from the hundreds of similar queries which have been addressed to us that many people suppose that they can obtain the current for lighting a room brilliantly and cheaply from one or two little batteries kept in a corner cupboard. We think it necessary to say at once that the use of primary batteries for electric lighting is at the best a method economically possible for only the smallest applications, such as night-lights, experimental work, scientific toys, etc. The battery for lighting any number of really serviceable lamps is an unwieldy affair, involving tremendous trouble and a quite unreasonable expense for upkeep. Ordinary Leclanché cells, whether wet or dry, are practically useless for the purpose ; possibly the least troublesome and expensive is the type that is fully described later in this chapter.

At the risk of repeating statements already made in earlier chapters, it is necessary to point out that electric energy is produced by primary batteries by chemical reaction in their cells effecting changes in the constituents of the material with which the batteries are charged. When these changes are completed the materials are used up, and chemical reaction and electric energy cease. The quantity of electric energy to be obtained from a battery is therefore limited by the quantity of material which can be changed. This quantity can be used in a short time by providing a channel of low resistance for the electric energy to flow through, or prolonged by having a channel of high resistance to the flow of the current. In the reactions which occur in the cells of a primary battery when the circuit is closed, gases are formed, and these must be absorbed or eliminated. Hydrogen is one of these gases, and is conveyed by the current to the negative element of the battery, where the hydrogen must be absorbed by some suitable material. If this absorption does not take place, the gas will "push back" the electric current and stop chemical reaction. The battery is then said to be polarised. Now, in a bichromate battery of the single-fluid type, polarisation soon happens if much current is drawn from the battery in a short time, as in electric lighting ; consequently this battery is unsuitable for electric lighting. A querist, asked us some time ago for the number of such cells required to light four 8-candle-power carbon filament lamps for a total period of 50 hours without recharging. Taking electric lamps that use from $3\frac{1}{2}$ to 4 watts for each candle-power of light, the four lamps of 8 c.p. equal 32 c.p., which would require from 111 to 128 watts. (Note that volts multiplied by amperes equals watts.) If 100-volt lamps were employed, it would be necessary to provide fifty bichromate cells in the battery to light them, since each cell gives an E.M.F. of 2 volts, and $50 \times 2 = 100$. As each lamp will take from 1·11 to 1·28 amperes of current, the four lamps in parallel will take from 4·44 to 5·12 amperes. As the ordinary pint bichromate cell polarises in fifteen minutes when more than $\frac{1}{2}$ ampere of current is drawn from it, the cells should be at least of 12-pint capacity, and these cells could not be charged with sufficient ingredients to furnish electric energy for even four hours.

Summing up, then, fifty 12-pint single-fluid bichromate cells, costing many, many pounds, would require to be recharged every three or four hours (for which operation about 1 cwt. of bichromate of potash would be required) to maintain four 8-candle-power lamps, each consuming from $3\frac{1}{2}$ to 4 watts per candle-power per hour. Current from a public main would do the work at a cost of, roughly, one penny per hour.

Better results may be expected if the double-fluid bichromate is chosen instead of the single-fluid type, since the depolariser is in a separate cell, and as a consequence the battery does not so rapidly polarise. But the number of cells would not be altered, and their size would have to be considerably enlarged to provide for a 5-ampere discharge during a period of fifty hours without recharging. The cost of a pair of cells fitted with zinc and carbon elements, suitable for a double-fluid bichromate battery of 2-gal. capacity would be 16s., and a battery of fifty cells would therefore cost £40. In any arrangement of cells the demand for current would be too great to be supplied for any reasonable length of time. By employing lamps of a lower voltage, fewer cells in series would be required; but the necessary volume of current, also the trouble from rapid polarisation, would be increased.

By the use of metallic filament lamps the number of cells required is greatly reduced. Osram metallic filament bulbs of 8 candle-power consume only 8 watts (1 ampere at 8 volts), and eight No. 1 Neotherm copper-oxide cells (possibly the primary battery that most nearly approaches the accumulator in its convenience of being easily recharged) connected in parallel, could maintain eight such lamps for between 100 and 150 hours without recharging. The soda hydrate for charging eight No. 1 cells of this type costs 4s. 8d.

The dry cell is of no use for electric lighting on any but the smallest scale. The E.M.F. of each cell does not exceed $1\frac{1}{2}$ volts, and therefore many cells in series are needed to light high-voltage lamps; and, as the internal resistance of each cell is very great, it is almost useless employing low-voltage lamps having a low resistance. Then again, as the depolarising substance in a dry cell (peroxide of manganese) yields its oxygen slowly, the elements soon polarise when a large current is being taken from the cell, and the light consequently gets dim, because current is kept back by increased resistance in the cell. Therefore high-voltage lamps, taking a small current, must be used, and a large number of rather large cells must be employed to light the lamps and keep them lighted. For example, to light a 25-volt 8-candle-power lamp for only an hour, it would be necessary to have seventeen dry cells in series, each cell being no less than 12 in. high and 8 in. in diameter.

Special Type of Lalande-Chaperon Battery

Coming down to small lamps, say 4-volt, of low candle-power, for use as night-lights, on models, etc., the use of a primary battery is more practicable ; and the most likely battery for such purposes is the Lalande-Chaperon, or Edison-Lalande, described in Chapter VI. The couple consists of zinc and copper-oxide plates in caustic alkali. The E.M.F. is low (·75 volt per cell), and therefore a 4·5 volt tension requires a battery of six cells. The $\frac{1}{2}$-volt margin is useful to overcome the resistance of connections and wiring. To be efficient, the cells must be large ; consequently, such an installation is cumbersome in proportion to its powers. Its non-portability, and the fact that it must not be disturbed when once set up, may in part account for its want of popularity. Nor is it a cheap battery ; cells of about the same capacity as that to be described are listed at 26s. each ; but these instructions will explain how they may be made for about one-fifth of that sum, or possibly less.

The advantages of the cell are that it is absolutely constant, yielding the whole of its initial charge in a steady and reliable current without polarisation ; in this respect it resembles an accumulator more nearly than most other primary batteries, not even excepting the Daniell. When at rest the elements suffer no change or deterioration, and may be left for long periods unattended. It is capable of a large output if required, the current being available on a heavy ampere rate, with a proportionately rapid exhaustion of the charge. It is probable that with large plates a discharge of 4 amperes to 5 amperes might be made without heating or disintegration of the negative material.

The cell may take a variety of forms, but that shown by Figs. 72 to 74 has been found both convenient and substantial. The general arrangement and the cellular construction of the negatives are of original design. The dimensions provide a theoretical capacity of at least 200 ampere-hours, and the internal resistance will probably be less than ·05 ohm, if the plates are spaced as directed and the electrolyte is of the strength given.

Fig. 72 is a sectional elevation on the line x x (Fig. 73), and Fig. 74 a cross section of the cover on the line Y Y. The construction of the cover is thus clearly shown. It is made up of two squares of well-seasoned wood $\frac{3}{8}$-in. thick, with the grain crossing ; the smaller fits easily within the cell, and the larger rests on its top. The squares are best dipped in pitch or asphalt before being nailed or screwed together while still warm ; but if preferred, brunswick black may be used to make the joint. Four pieces of hard wood are attached, to take the tangs of the several elements, in the positions shown. The longer pair of these pieces measure $\frac{3}{4}$ in. by $\frac{3}{4}$ in. by $2\frac{1}{4}$ in., and the central ones are $\frac{3}{4}$ in. high by $1\frac{1}{2}$ in. long ; but they differ in width, one being $\frac{3}{4}$ in. and the other $\frac{7}{8}$ in. All are secured to the cover with suitable screws, the joints being made with pitch or brunswick black. A $\frac{3}{4}$-in. hole is bored in each cover, as in Fig. 73.

The tangs of the elements are brought through slots in the cover, well cleaned, and firmly clamped in position with bolts and wood screws. The three

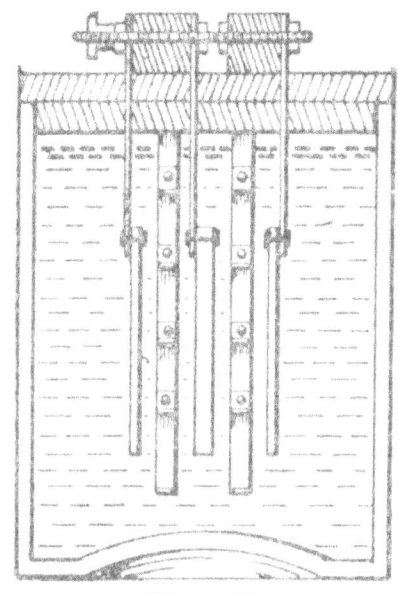

Fig. 72.

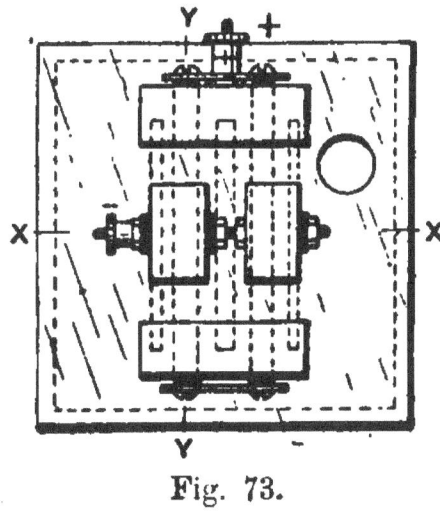

Fig. 73.

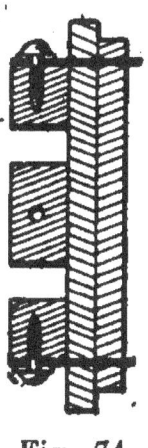

Fig. 74.

Figs. 72 and 73.—Vertical Section and Plan of Battery for Electric Lighting. Fig. 74.—Section of Cover.

zincs are coupled by a bolt and four lock-nuts, and the coppers are bridged by copper straps on both sides (see Figs. 73 and 74), and secured with mushroom-headed wood screws; terminals are provided as shown.

The negative plates (*see* Fig. 75) are built up of sheet copper in the grid form, as used in accumulators. By this means not only is the paste well secured, but also the ohmic resistance of the plate as a whole is reduced, and the employment of a gauze facing is dispensed with. In the original form the oxide plate was moulded in one piece, and held in a rebated frame by bolts; in another the paste was enclosed in a copper-gauze envelope (*see* p. 79). The grid type, however, is most reliable for a large cell, and it may be cleared out and re-pasted when the oxide is reduced.

The framing consists of one strip of stout sheet copper (No. 18 or No. 20 gauge), 1 ft. 5 in. long and $\frac{1}{4}$ in. wide. This is bent at right angles at a distance of $6\frac{1}{2}$ in. from each end, leaving a 4 in. space between the tangs, which are bridged by a $4\frac{1}{2}$-in. strip of the same gauge bent up $\frac{1}{4}$ in. each end, and riveted each side to form a 4-in. square and two $2\frac{1}{2}$-in. suspenders. Six similar $4\frac{1}{2}$-in. strips, bent in the same manner as the first, are required to complete the grid. Each is cut half through with a hack saw at 1-in. intervals, to interlock with the others at the points of crossing. This is best done by clamping each set together in the vice, and slotting the six at one operation. They are then assembled in the frame, and secured to it with two copper rivets apiece; no solder is permissible. If desired, these inner strips may be of lighter gauge than the main framing. It is not necessary to give an over-hang to the grid apertures (as in the case of accumulators), and the paste is retained well enough by the simple squares if the battery is not disturbed or submitted to an excessive rate of discharge.

The two outer zincs may be 3 in. by 3 in. by $\frac{1}{8}$ in., and the middle one (Fig. 76) of the same size by $\frac{1}{4}$ in. thick. The zincs are supported by stout copper straps riveted and soldered to them. The junction must be carefully washed free from flux, dried, and thoroughly coated with pitch to protect it from local action, which would rapidly destroy the connection and ruin the cell. The plates are spaced $\frac{3}{8}$ in. apart. Heavier zincs or thicker grids are not required in this cell, because the available quantity of electrolyte

Fig. 75.—Negative Grid.

Fig. 76.—Central Zinc Plate.

Fig. 75.

Fig. 76.

($\frac{1}{2}$ gal.) will probably be exhausted before the elements, and it would be a waste of time to thoroughly clean up and discharge with alkali, only to replace half-consumed zincs and partially converted negative plates.

In making the battery, the grids should be made and pasted in the first place, so that they may have

I

ample time to dry thoroughly while other parts are being made. The paste is composed of 9 parts (by measure) of black oxide of copper (CuO) and 1 part of plaster-of-Paris (best quality). Test the latter before use ; if it does not set hard within a few minutes it is useless for this purpose, as the very small proportion used as a binder for the oxide must be of the best. Mix the oxide and the plaster in the dry state very thoroughly, then add just sufficient cold water to work up into a very stiff paste.

Attention to this detail is especially requested. If the mixture is made in a basin with a spoon in the first place, let the paste be only crumblingly moist, then transfer it to a glass or slate slab, and work up briskly with a palette knife. When an even consistency is obtained, lay the grid on a smooth, flat surface, and force the paste vigorously into every corner of all the sixteen compartments ; fill them well up and smooth off. Now reverse the plate (by sliding it from its support), and attend to the opposite side, filling in where necessary and smoothing off. Leave some excess of mixture standing slightly above the squares, and weigh heavily on it with another slab to compress the oxide. Finally level off with the palette knife, and stand aside to dry.

To be successful, the whole operation must be quickly and deftly carried out. Moisten only sufficient mixture at one time to fill one grid. Use the least possible quantity of water to work up into a very stiff paste of even consistency throughout. From the first addition of water to the final smoothing off, the time occupied should not exceed two minutes. The paste must dry thoroughly ; three days in a warm, dry, and airy situation may suffice, but a week is not too long. Do not on any account heat the grids to hasten the drying, or the material will pulverise.

For the glass containers half-gallon pickle-jars or onion jars may be used. Those of greenish shade should be selected; white glass, containing lead oxide, may be affected by the caustic alkali. All must now be cut off at the shoulder with a diamond glass-cutter. They will then measure $6\frac{1}{4}$ in. high by 5 in. square, and are admirably adapted to their purpose. The sharp edges should be removed with an old file or a scrap of gritstone.

If the covers and the tangs of the elements are made as directed, the latter will be suspended in the jars as in Fig. 72. The negative plates must be fully submerged, the solution covering their tops by at least $\frac{1}{2}$ in. The two heavily dotted lines near the cover indicate a layer of oil floating on the surface of the electrolyte to exclude air; there must be a small air space between the oil and the cover. The tangs must be thoroughly coated with asphalt.

When setting up, it is important that the right quantity of electrolyte is poured into the jars at one operation. Therefore measure the capacity of each cell with plain water, and scratch a line on it at the correct level for future reference. The displacement of the elements must, of course, be allowed for. To do this, cut five pieces of thin wood to the dimensions of the coppers and zincs respectively, bind them together, and use the bundle to represent the elements when gauging the several levels of the jars. Do not plunge the elements themselves (when finished and attached to their covers) into water, as the dried oxide plates must be first moistened and saturated with the solution itself only.

All being complete, set the jars in their permanent position on a shelf where they will not be subjected to extremes of temperature, dust, or excessive damp-ness. Fill up to the gauge marks with electrolyte, avoiding splashing; carefully wipe off any trace of solution from the edges of the cells, and see that they are clean and dry outside. Then steadily lower each set of plates into its jar without unnecessary dis-turbance of the liquor. All being in position, insert a small funnel in the hole in each lid in succession, and pour slowly on to the surface of the potash a couple of ounces, more or less, of vaseline oil until it attains a depth of about ¼ in. above the electrolyte. Cork the lid apertures, to exclude dust, and do not again disturb the battery. A small quantity of paste may disengage itself from the surface of the negatives, and settle at the bottoms of the jars; but this is of no consequence.

Now make the connections with guttapercha-covered wire of large gauge (No. 14 or No. 16 s.w.G.). Leave the copper terminal of the first cell free, and connect its zinc to the copper of the next, and so on throughout the series, leaving the zinc of the sixth cell free. The free terminals of the first copper and the last zinc are, of course, connected to the line wires. Be sure to avoid short-circuiting the cells collectively, or any one of them individually; that is, do not couple the terminals with a short wire of little or no resistance, or serious damage may result. The safety of the battery is insured by inserting a fuse in the main circuit; one to blow at 2 amperes will suit the purpose. The protection thus afforded will save the cells from heating and disintegration in case of a "short" occurring in the lines at any time.

The electrolyte is made up of 8 oz. of caustic potash (potassium hydrate) to every quart (40 oz.) of distilled or rainwater. If neither is procurable, main water may be used; but it must be briskly boiled for at least half an hour, and then allowed to cool to about 120° F., or just too hot to handle. When dissolving, stir briskly to hasten solution, and do not spill the liquid on the skin or elsewhere, as it is destructive to many substances. Do not pour into the jars while still quite hot; but once the solution is made, make reasonable haste to set up the battery and to exclude air from the electrolyte by means of the oil layer. If kept too long exposed to the air, the hydrate will slowly deteriorate by absorption of carbonic acid ; a trace of warmth, however, will not injure the negatives.

Allow one hour's saturation of the plates before taking current from the cells. Then run a 4-volt 0·5-ampere lamp for an hour or two, which will have the effect of consolidating the grid paste to some extent. An adjustable resistance is advisable to save the possibility of over-running the lamp. Use No. 16 for line wires, and exercise care in their perfect insulation.

The battery will be found most reliable if used well within its capacity ; its maximum output should not exceed 2 amperes if economy is desired. Four 0·5-ampere 4-volt Osram lamps (in parallel) may be run at the same time, or two 1-ampere 4-candle-power bulbs of the same voltage ; but if such a comparatively large demand is made on it, the initial charge will be exhausted at a proportionately rapid rate.

Electro-plating or the driving of small electric motors (such as those applied to phonographs) is also within the capacity of the cells. A note as to its capacity for accumulator charging is given on p. 140.

CHAPTER XII

PRIMARY BATTERIES FOR ACCUMULATOR CHARGING

WITH the popularity of the petrol motor, in which the explosive charge is ignited by electricity frequently generated by accumulators, and with the general movement in favour of replacing primary batteries by accumulators in electric bell and telephone installations, in electro-plating, etc., there has arisen a considerable industry in charging these appliances. In all towns nowadays someone or other has laid down the necessary plant to cope with this lucrative and constantly growing business. The necessary current is taken from the town main, where there is any, or from a dynamo, when there is no public supply, and accumulators can be charged from either of these sources far more cheaply than by any other means. There are, however, in out-of-the-way parts of the country many thousands of persons anxious to avail themselves of the big advantages possessed by the accumulator, but who are at a loss as to how they may overcome the particular disadvantage that is inseparable from its use, namely, the necessity of re-charging it at fairly frequent intervals. The difficulty is easily got over by laying down an engine and dynamo, but this course is often too expensive. Sometimes it answers to keep a spare accumulator, so that one may be taken to the nearest town for charging while the other is in use. There is, however, a large number of people who have attempted to overcome the difficulty in yet another way—that of using large primary batteries for charging the accumulators. The method has obvious advantages and disadvantages. It renders the user independent of the town charging plant, but it involves much necessary work and is relatively expensive.

The idea of employing ordinary bell batteries, etc., for accumulator charging need not be encouraged, because their use is quite hopeless. A special battery must be provided.

In this chapter are described the three best batteries for the purpose, namely, special forms of the bichromate, Lalande-Chaperon, and Daniell batteries. The choice between these rests largely upon one consideration—the ease with which the necessary charging materials may be obtained. Sulphate of copper, zinc, and sheet copper all being commonly obtainable in the nearest town of any size, the choice would probably fall on a modification of the Daniell battery as the most convenient type to adopt. It is slow in charging, however, and necessitates, 'as a rule, a spare accumulator, which can be taken into use while the other is charging. The bichromate or chromic acid cell, while being a much more powerful battery, has the objection that the requisite exciting salts for the electrolyte are not always readily procurable, and the same may be said for the Lalande-Chaperon.

Accumulators Described Elsewhere

Not much will here be said with regard to the nature of an accumulator and to the methods of charging; the subject is so large that it has been thought better to devote a separate handbook to it, and therefore in " Electric Accumulators," which is uniform with the present book in size and price, the inquirer can read for himself everything that is necessary on the part of an accumulator user to know.

Elmwood Battery

One of the best primary batteries for accumulator charging is the Elmwood, which in all essentials is the same as the star zinc and crow-foot patterns of the Daniell battery, which are described on p. 71. The Elmwood can be made up at small cost, as shown by Fig. 77, in which A is a heavy zinc casting fixed in any convenient manner (a lug, as in Fig. 44, or a tripod, as in Fig. 46, answers the purpose) to the cover of the glass containing cell B; c is a bent copper plate having a stout insulated copper wire D proceeding to the outside to form the other terminal. This copper sheet should extend about one-third up the cell. The jar is filled to within an inch of the top with a saturated solution of sulphate of copper; there should also be a good thick layer of

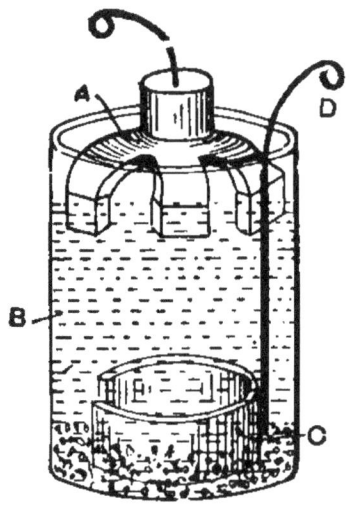

Fig. 77.—Elmwood Cell.

crystals left at the bottom of the jar to keep up the strength of the solution. The working E.M.F. of this cell is rather low, but it is very constant and reliable. About four or five cells will be required coupled in series to charge a 2-volt accumulator. The exact size of the various parts is immaterial, and may be proportioned to the available material. For instance, 4-lb. glass jam jars may be utilised for the outer vessels. The zincs can be bought or cast at home in the claw form in a stiff clay mould from a wooden pattern, and would weigh between 2 lb. and 3 lb. each when new. It is only the zincs that are consumed, and when the part that is immersed in the solution is eaten away, the remainder can, together with fresh zinc, be melted up and cast into a fresh supply of zincs so as to avoid waste. The heavier the zincs, of course, the less frequent will be the renewals. No. 20 sheet copper will do for the other element, which should have a 14 gauge or 16 gauge leading-in wire soldered firmly to it to form one terminal. It is not necessary to remove either of the elements when the battery is out of use, as no local action takes place.

Double-fluid Bichromate Battery

The next battery to be described is a special design of the granule type—a double fluid bichromate, with a potassium bichromate and nitrate solution in the outer cell and brine in the inner one. The cell has a low internal resistance, and an E.M.F. of 1·8 volts.

For a small output, a round salt jar makes a good outer cell, but for large outputs the cells are preferably made rectangular, and the outer containing cell of wood made water- and acid-proof. A good size for medium outputs would be 9 in. high, 9 in. long, and 4 in. deep, and for large discharges 12 in. by 12 in. by 5 in. or 6 in. Each cell contains a rectangular porous pot, in which is the zinc plate. Between the porous pot and the outer cell are two plates of carbon, made up of several arc-light carbons connected together at one end by a lead capping, the remainder of the space being filled with crushed gas-retort carbon.

Fig. 78 is a section of a rectangular cell, and Fig. 79 the plan, in which A is the outer cell of wood, made water- and acid-proof by coating the inside with marine glue of any acid-proof cement. B is the porous pot a little higher than the outer cell, and of such a width as to occupy a little less than one-third of the inside of the cell A. C is the zinc plate, $\frac{1}{4}$ in. thick, of sufficient size to slide loosely in the porous pot and

project about ¾ in. to 1 in. above ; at the top edge
at one corner is fitted the terminal D. With the
common salt solution used in the porous pot it is
not necessary for the zinc to be amalgamated.

The carbons E consist of arc-lamp carbons built
into a plate by placing them side by side about ⅛ in.
apart, and binding them together by running molten
lead round the top to form a capping as at F. The
two plates are connected together by two lead straps
G by "burning," but this can remain until the cell is
built up. H is the terminal which is secured to the

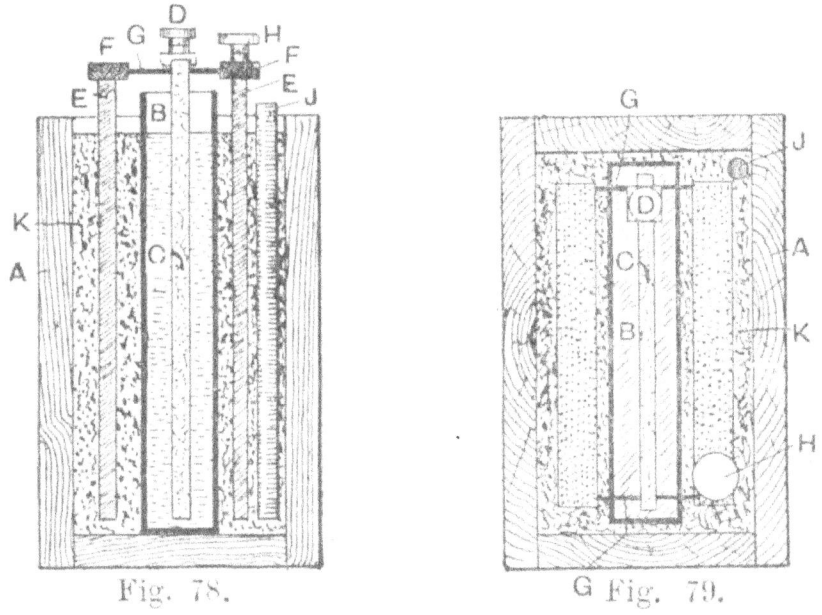

Figs. 78 and 79.—Vertical Section and Plan of Special
Type of Double-fluid Bichromate Cell.

lead capping at one end, and J is a glass tube 1 in. in
diameter, open at both ends, and of such a length as
to reach about 1 in. above the outer cell and nearly
to the bottom. The object of this tube is to allow the
exhausted solution to be withdrawn by a glass syringe.

The cell is put together by placing the porous
pot in position, the carbons are centred in the space
between the porous pot and the outer cell, and the
space is filled with crushed gas-retort carbon K, which
should be graded in size from ¼ in. to ⅛ in., and all the

dust carefully removed. The zinc plate is now put in the porous pot, and the lead straps G "burned" (fused) to the lead capping of the carbons. To prevent the lead corroding, it is coated with acid-proof varnish.

For the outer or carbon cell, make a saturated solution of bichromate of potash with cold water (three-quarters of an hour will be sufficient to make the solution strong enough if stirred two or three times), allow to settle, and pour off the clear solution. In a glazed earthenware pan, mix, in sufficient quantity to fill the cells, 2 lb. of the above with 2 lb. of sulphuric acid (slowly adding the latter to the former), and dissolve in this 2 lb. of nitrate of potash. If all the nitrate is not dissolved after some time, bichromate solution is added a little at a time, continually stirring. Approximately, $2\frac{1}{2}$ pt. of solution will be required for each of the smaller cells described, and the above quantities will make about 3 pt.

For the inner cell make a solution of common salt, and to prevent the solutions mixing by percolation through the porous pot, the salt solution should be made of the same density as the bichromate solution ; this can be ascertained by means of a hydrometer. The exhausted solutions are removed by means of a glass syringe, thus avoiding taking the battery apart each time of recharging.

For charging a 4-volt accumulator, a battery of three such cells, medium-sized, joined in series, will be required. The carbon will be the positive, and the zinc the negative, pole of the battery. Join the carbon to the positive ($+$) terminal of the accumulator, and the negative pole of one to the negative of the other. As this battery will give off acid fumes, it should be set up in an outhouse or other place having ample ventilation.

Lalande-Chaperon Battery

Finally, the Lalande-Chaperon cell, which is fully dealt with in Chapters VI. and XI., may be used.

In the former chapter is described a pattern especially adapted for this purpose, but there is a difference of opinion as to its particular suitability for the work. However, there can be no doubt that, although the voltage is low (0·7 volt per cell), there is no reason why the battery should not give fair service for a time if properly constructed. The copper wires connecting the battery to the accumulator should not be thinner than No. 18 or No. 16 gauge. The two copper elements in each battery are, of course, to be connected together, forming one pole, while the zinc forms the other ; and the four cells require connecting in " series "—that is, the zinc of one battery to the copper of the next, and so on, leaving one free copper and one zinc at each end for connection to the accumulator. The copper of the battery requires connecting to the positive terminal of the accumulator and the zinc to the negative terminal when charging.

Four cells of the battery described on pp. 125 to 133 of the preceding chapter will charge small 2-volt accumulators for hand lamps or ignition purposes, but seven cells would be required to charge efficiently a 4-volt storage battery. For this purpose an ammeter and resistance should be included in the circuit of the battery and the accumulator to adjust the current to the charging rate of the latter.

CHAPTER XIII

MEDICAL BATTERIES

MEDICAL batteries have their legitimate and proper employment when adopted under the advice of a qualified doctor. Unfortunately, medical electricity has been extensively exploited at the expense of the public's money and self-respect, and there still linger in very many quarters much doubt and suspicion where the medical use of electrical appliances is concerned. However, subject to a doctor's approval, the electric battery can be advantageously used in the treatment of sciatica and similar ailments.

There are two kinds of galvanic batteries in use for medical purposes.

One is the well-known Leclanché battery, made up of a number of small cells, in a box with a switch to take in or cut out cells as required. These, when purchased ready made, are sealed, and are not easily repaired or refilled at home. When home-made, the tops may be left unsealed, and the cells can be easily refilled. Dry cells are merely a variety of the Leclanché series, and are not easily refilled.

The next kind is the chloride of silver battery, also made up of a number of cells. This is more powerful than the Leclanché, and may be used for cautery purposes. When home-made, this battery can be easily repaired.

One form of chloride of silver dry cell for medical use (patented in 1898) is shown in section by Fig. 80, in which A is an outer cylindrical case of metal, recessed at the top to receive a hard fibre cap B carrying two long brass terminals, which pass through a battery panel M, on which the necessary series or parallel cell connections are made by means of brass straps. C is a glass container shaped as shown; D is a zinc plate or rod; E is a chloride of silver rod, immersed in a solution of zinc sulphate, and connected to the terminals with rubber-covered copper wires. F are hard fibre separators. The cell is sealed as follows : G is a mixture of resin and guttapercha poured in hot, becoming hard

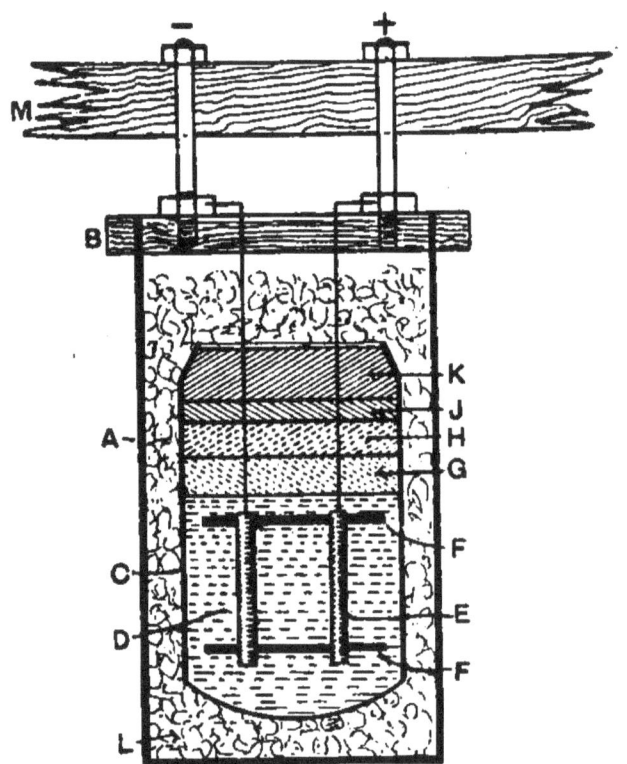

Fig. 80.—Section of Silver-chloride Medical Cell.

when cool; H is resin dissolved in resin oil, and remaining semi-plastic; J is a thin layer of plaster-of-Paris, which is allowed to set hard before the final sealing K, also of plaster-of-Paris, is poured in. An absorbent packing L surrounds the glass container. The E.M.F. of each cell is probably 1·3 to 1·4 volts.

A cheap form of medical battery of small cells is made with chemical test-tube cells, each cell containing a small carbon rod that is enveloped in blotting paper which has been soaked in a strong solution of sal-ammoniac ; the rod is packed tightly in a tube of sheet zinc, which is then enveloped in the wet blotting paper and packed loosely in the test-tube. The ends of the carbon rods are coppered (*see* p. 20) and soldered to wires, the other ends of which should be soldered to the zinc tubes, the carbon of one cell being soldered to the zinc tube of the next cell. The cells thus prepared may be packed in a box with sawdust or paper. Terminals on the side of the box connect the free copper at one end and the free zinc at the other end of the battery.

The chloride of silver battery may take the form of a large number of test-tube cells arranged in series to provide sufficient E.M.F. to overcome the high resistance of the human body. As the number required will depend on the desired strength of current, which will vary with circumstances, the decision must be left to the medical attendant.

Each small cell may be constructed as follows : Obtain some sheet zinc, cut off a strip wide enough to make a small cylinder fitting the containing tube, and long enough to match the length of tube, with about 1 in. over to form a connecting tang. Also obtain some No. 18 gauge silver wire, and cut from it a piece 1 in. longer than the tube. Dissolve some scrap silver in dilute nitric acid, or get some nitrate of silver and dissolve it in distilled water ; then add brine to this until it ceases to throw down white curds. These curds are chloride of silver, and must be separated from the brine by pouring all the liquid portion away. Some of these curds must then be carefully dried, then heated on a piece of porcelain until they fuse, and three-fourths of the silver wire rolled in the fused chloride of silver until thickly coated with it. They are now said to be coated with " horn " silver. Some of the remaining curds may next be spread as a paste on strips of thick blotting paper and rolled round the coated silver wire, then bound thereto by several turns of soft cotton. This porous paper will then serve the purpose of a porous cell. The upper fourth of the zinc cylinder should be coated with asphalt varnish to prevent chemical action on this part. The cylinder is then put into the containing tube, the swaddled silver wire is placed centrally in the zinc cylinder, and the space filled with a solution of caustic soda. The cells, thus prepared and charged, must be packed in an upright position in a wooden box with paper or sawdust between them, and connected in series by attaching the zinc of one cell to the silver wire of the next, and so on throughout the whole number. The current will be taken off from the silver wire at one end of the series, connection being made to the zinc cylinder at the other end.

CHAPTER XIV

COLLECTOR SWITCHES AND SWITCH-BOARDS

A Simple Collector Switch.—A collector switch of simple design may be made as follows : On a piece of good mahogany or other hard wood $\frac{5}{8}$ in. thick describe a circle 4 in. in diameter. On the circle mark off twenty-seven points $\frac{3}{8}$ in. apart from each other ; and at each point drill a hole to take a brass pin, made of $\frac{1}{8}$-in. or $\frac{5}{32}$-in. brass wire and $1\frac{1}{4}$ in. long. The arrangement is shown in Fig. 81. The pins

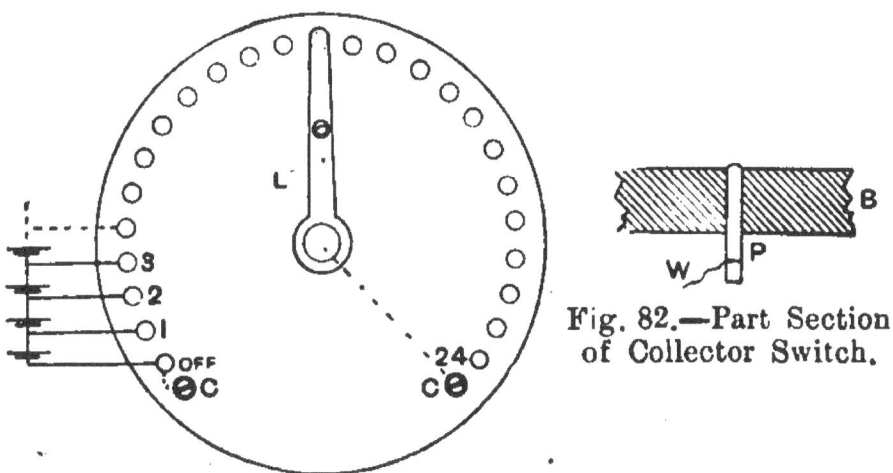

Fig. 82.—Part Section of Collector Switch.

Fig. 81.—Collector Switch.

should be a fairly tight fit in the holes, and barely project above the level of the baseboard. The two screw terminals c are for connection to the coil, the first stud, OFF position, being also joined to the left-hand terminal, the switch lever L being joined to the right-hand terminal. The lever L, of spring brass, should be $\frac{1}{4}$ in. wide at the point where connection is made to the pins. Solder the covered wire leads from the cells to the bottom of the pins as in Fig. 82, where B is the baseboard, P the pin, and w

J

the soldered wire connection. The connections to the cells are shown in Fig. 81. The cells could be packed in a box, the collector switch being fitted on the box lid.

This switch gives every facility for increasing the number of cells as required. It may, for example, be conveniently used in conjunction with a 24-cell medical battery. A box is made to hold the cells, the collector switch fitted on top of the lid, and the cells wired to the collector switch, as shown in Fig. 81. The massage electrodes may then be connected to the terminals marked c, and by moving the switch lever any number of cells in series from 1 to 24 may be used. If desired, the number of contact studs could be reduced so as to increase the battery by steps of three cells, but the more usual practice is to provide a switch by means of which single cells may be added as required.

In Fig. 81, each cell of the battery is represented thus ‖‖, the short thick line denoting the zinc terminal, and the long thin line the carbon terminal. If the cells are connected to the studs as shown, the positive current will pass from the terminal marked c via the switch lever. The " off " terminal will therefore be connected to the outer zinc (negative) of the battery.

Switch-board for Primary Cells.—Fig. 83 shows the connections for a switch-board suitable for twelve cells, the connections allowing of practically any combination of the cells being made. The switch-board is arranged to be fitted with both ammeter and voltmeter. If desired, the battery leads may be connected to screw terminals, the operator using brass straps or bare wire to make the necessary connections. If in Fig. 83 the two inner rows lettered A, B, C, to U be deleted, the remaining portion of the figure will show the wiring for screw terminals, two single-pole electric light switches to replace the ammeter and voltmeter terminals. For ease of manipulation, what

is known as the " link and test-hole " system is to be
preferred, in which case Fig. 83 shows the wiring, etc.
Brass test-holes, with nuts, may be obtained in
several sizes, and at the prices of screw terminals.
Or if a lathe or drilling machine is available, ⅜-in.
brass bolts 1¼ in. long may be used if a hole ⅛ in. in
diameter is drilled centrally lengthwise, as shown in
Fig. 84, a number of links also being made of ⅛-in.
brass wire, shaped as shown in Fig. 85.

The board should be of sound hardwood ⅝ in.
thick. Mark out the positions of the test-holes, the

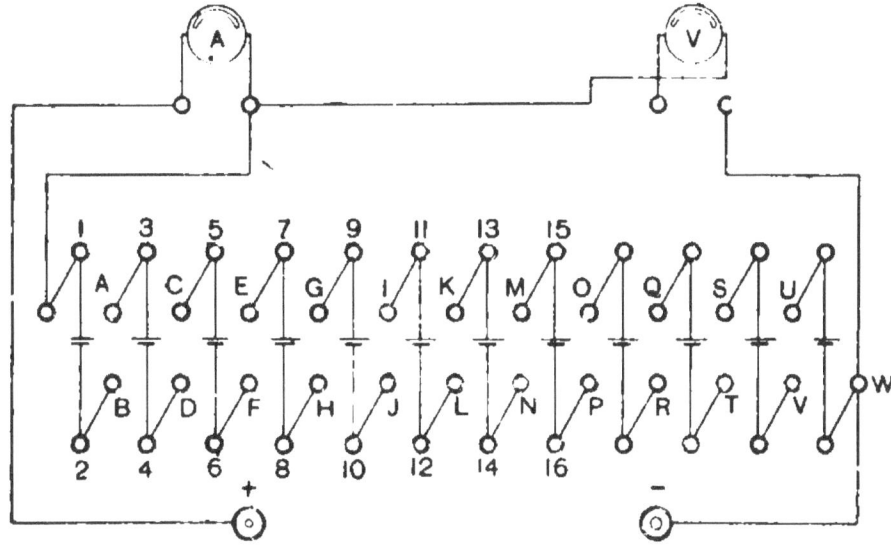

Fig. 83.—Switch-board for Twelve Cells.

centres being 1 in. apart. Then drill holes in the board
to receive the test-holes, which should have a moder-
ately tight fit. Two large size screw terminals are
shown at the bottom of the board in Fig. 83. A length
of flexible wire fitted with ⅛-in. brass wire ends will
be required (see Fig. 86). Use No. 18 s.w.g. copper
wire, cotton and rubber covered, for the battery leads
to the test-holes, and No. 10 s.w.g. copper wire for
the wiring of the ammeter, voltmeter, and the two
screw terminals, soldering the wires to the terminals,
all the wiring shown in Fig. 83 to be made at the back
of the board.

To join the cells in series, insert links in test-holes
A B, C D, E F, and so on. Should four cells in series
be required, link up A B, C D, and E F, and join H to W
with the flexible cord. To join cells in parallel, insert
links in A 1, B 4, C 3, D 6, and so on. Say three cells
in parallel are required, then link up A 1, B 4, C 3, D 6,
and join F to W with the flexible cord. Let a combination of two in parallel four in series be required,
then link up A 1, B 4, D C, E 5, F 8, G H, I 9, J 12, L K,
M 13, N 16, and join P to W.

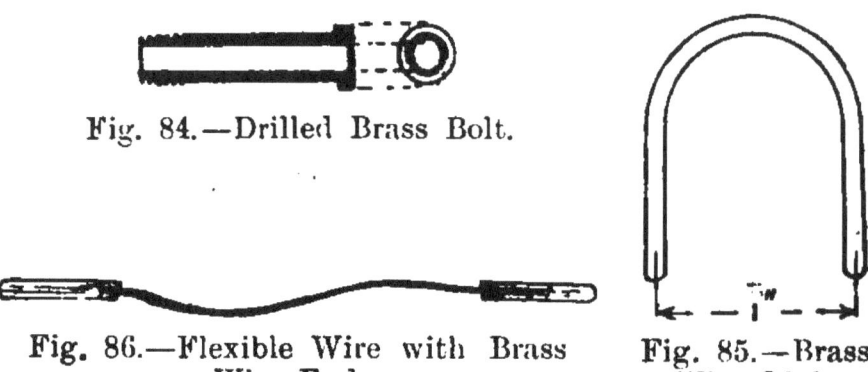

Fig. 84.—Drilled Brass Bolt.

Fig. 86.—Flexible Wire with Brass
Wire Ends.

Fig. 85.—Brass
Wire Link.

Any combination possible with twelve cells can
be made on the board. It will be seen that the series
links are inserted vertically in the two inner rows,
and that the parallel links are inserted diagonally in
the two upper and the two lower rows of test-holes.
Normally the voltmeter test-holes should be free, the
instrument being brought in by inserting a link in
the test-holes. The ammeter should normally be
short-circuited with a link, the instrument being
brought in by withdrawing the link. In wiring the
ammeter and voltmeter, care should be taken to see
that the positive lead is connected to the positive
terminal of each instrument.

CHAPTER XV

MAKING AN ELECTRIC TORCH

THE arrangement of the interior parts and fittings of an electric torch about 7 in. long are clearly illustrated by Fig. 87. The whole series of cells and their fittings are shown loose for clearness, but they should be made to fit close in the outer case, or else packed tight with paper to prevent movement and consequent chafing.

The little 4-volt lamp A is placed in a recess at the top, the recess being formed by the reflector B and the lens C. It is necessary to mount the lamp in a recess of this kind to protect the bulb from injury, and to magnify the small light given by a 4-volt carbon-filament lamp, this being the largest lamp practicable in a torch 7 in. long. A tiny metal-filament lamp will give more light.

The lamp is screwed into a disc of ebonite made to fit tight into a small brass case holding the lens and the reflector, this case fitting tight into the end of the outer case, which may be made of ebonite, papier-maché, or stout cardboard. One terminal of the lamp is connected to the contact spring D, made of thin hard sheet-brass or german silver and attached with small screws to the under side of the ebonite disc, whilst the other terminal is connected to an insulated conductor passing down the outside of the case to a stud as shown. This conductor may be a short piece of No. 22 covered copper wire, or a thin strip of sheet-copper varnished to match the colour of the case. The electrical circuit is completed through this stud and the end of the spring lever E when the latter is pressed by the hand against the stud, then by the connection of this lever with a contact spring F inside the case, making contact with the zinc cylinder of the lowest cell of the battery.

When all the cells are pushed up tight to each other, the carbon cap of the top cell will be in contact with the spring D, the next cell with G, and the carbon cap of the lowest cell with H; and as these are all german silver springs soldered to the bottoms of the zinc cases of the cells, the circuit will then be complete.

The cells are of the dry type, as used in flash lamps, and are fully described in Chapter VIII. They must be made to fit the space left for them in the outer case. If the outer case is 7 in. long and $1\frac{3}{4}$ in. in internal diameter, the zinc cases of the cells should be only $\frac{1}{16}$ in. less in diameter; and if the lamp with its reflector occupies 1 in. of the length, the remaining 6 in. must be divided into three equal compartments, and the length of the cells arranged to fit these, after making allowance for the caps of the carbons and the contact springs.

The cell cases are made of thin sheet-zinc, each being in the form of a little cup made water-tight in all its joints. A small disc of cardboard is pasted into each bottom, for the inner cell to stand on, the insides lined with blotting-paper pasted on with flour paste, and the german silver contact springs G and H soldered on the outside bottom of the upper and middle cells.

The inner cells are small bags of canvas with a disc of porcelain or glass in the bottom of each bag. On this disc will stand the bottom end of a carbon rod, any thickness, and just long enough to fit the cell, with a cap of brass made to fit the top as shown. These carbon rods are packed tight in the canvas cells with a pasty mixture of finely powdered peroxide of manganese 5 parts, finely powdered carbon 5 parts, zinc chloride 1 part, sal-ammoniac in fine powder 1 part, glycerine $\frac{1}{2}$ part, water 1 part, all well mixed; this should be put in a little at a time, and tightly rammed down with a small wooden rammer. The top of this composition should be coated with a

mixture of pitch, resin, and paraffin' wax in equal
parts, melted together, and poured on in a thin layer
while hot. When this has cooled, it must be pierced
with a red-hot wire in two or more places to form

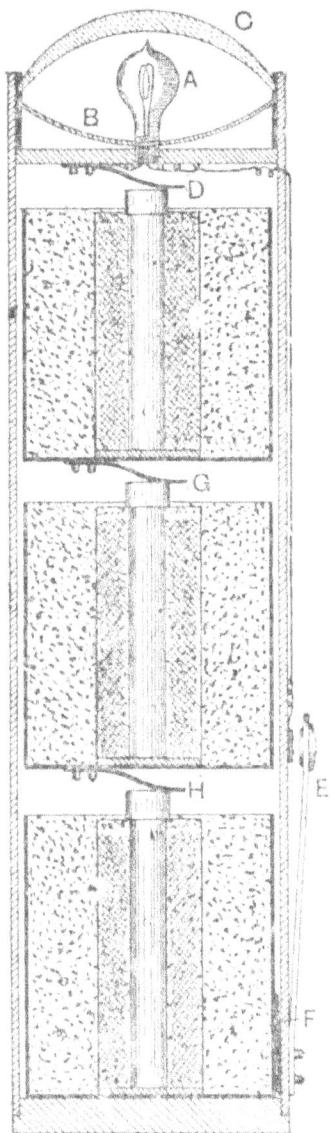

Fig. 87. – Section of Electric Torch showing Arrangement
of Cells and Fittings.

vent holes for the gas generated in the cell while
working. The inner cells, thus prepared, are then put
in the centre of each zinc cell, these zinc cells being
filled with a mixture composed of flour paste 1 teaspoonful, glycerine ½ oz., zinc chloride ¼ oz., and

sal-ammoniac ½ oz., boiled in water 4 oz. ; this should be poured in whilst hot, and allowed to cool, then covered with the mixture used for coating the inner cells.

As a 4-volt lamp is to be used, and each cell has an E.M.F. of 1·4 volts, three cells must be employed to get the required voltage. One cell 6 in. long would not provide enough volts, neither would two cells 3 in. long, since the combined voltage would be only 2·8 volts. But if a longer tube could be used for the torch, the cells could be of greater length, and they would then last longer than small cells. When the cells are exhausted, they must be refilled, as there is no other way of recharging them.

CHAPTER XVI

POLE FINDERS

WHEN an electric current is passed through a neutral solution containing some salt of potassium or sodium, the metal is liberated at the negative electrode. This (potassium or sodium) at once reacts with the water, producing hydrogen gas and potassium or sodium hydrate.

The reaction may be shown thus: K_2SO_4 (potassium sulphate) breaks up to K_2 and SO_4. The K_2 with the water forms KHO and H. The SO_4 with

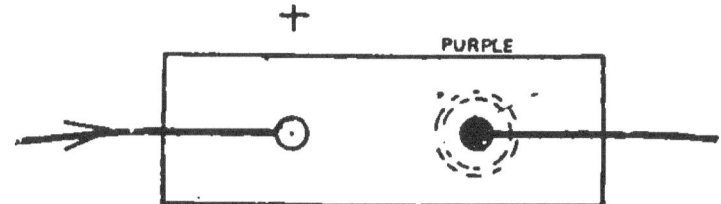

Fig. 88.—Negative Pole Indicator.

the water forms H_2SO_4 and O. Thus it is that oxygen gas appears at one electrode and hydrogen at the other.

Now the presence of the alkali at the negative electrode may be demonstrated by having in the solution some chemical indicator, for example, phenol phthalein. This, with neutral or acid solutions, is colourless, but with alkaline liquids is of a reddish-purple colour. Hence a reddish-purple liquid round the negative electrode would show the formation of the alkali. On, however, shaking the liquid, the acid formed at the positive electrode, and the alkali formed at the negative electrode, in equivalent quantities, re-combine; a neutral solution is produced and the colour is discharged.

Either of the following solutions may be made up. (A) In 1 oz. of distilled water dissolve 2 or 3 grains of sodium sulphate, and add a few drops of a 1 per cent. solution of phenol phthalein dissolved in alcohol. Normally this solution is colourless. On passing current through the solution a reddish-purple colour will be observed at the electrode connected to the negative power lead (see Fig. 88). (B) Instead of the phenol phthalein, add to the sodium sulphate solution one drop of a 1 per cent. solution of methyl

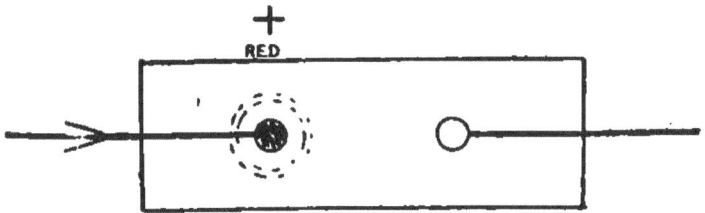

Fig. 89.—Positive Pole Indicator.

orange dissolved in alcohol. Normally the solution is of a yellow colour in neutral or alkaline solutions, and red in acid solutions. Therefore, on passing current through the solution, the presence of the acid at the positive electrode is indicated by a red colour at the electrode connected to the positive power lead (see Fig. 89).

These solutions are very sensitive. The (A) solution, being colourless normally, is to be preferred. There should be no difficulty in getting either solution made up by an analytical chemist. The electrodes may be of copper, 1 in. apart.

Pole-finding paper can also be used. Made as below, it needs to be moistened and touched with the ends of the two wires, which should be about $\frac{1}{4}$-in. apart, the positive pole producing a brown spot. Mix 1 oz. of best starch with distilled water to form a thick creamy paste and stir in boiling water until the starch becomes translucent. Add the greater part of a pint of water in which has been dissolved $\frac{1}{2}$-oz. of potassium iodide and 1 oz. of potassium nitrate. When cool, immerse in it pieces of white filter paper and dry in the dark.

CHAPTER XVII

THERMO-ELECTRIC BATTERIES OR THERMOPILES

THE means of producing an electric current direct from heat energy is known as a thermopile (from *thermos*, warm, and *pile*, battery). When many short bars of dissimilar metals are fused together at their ends to form a zigzag line, as shown in Fig. 90, and one set of junctions is made hot whilst the opposite

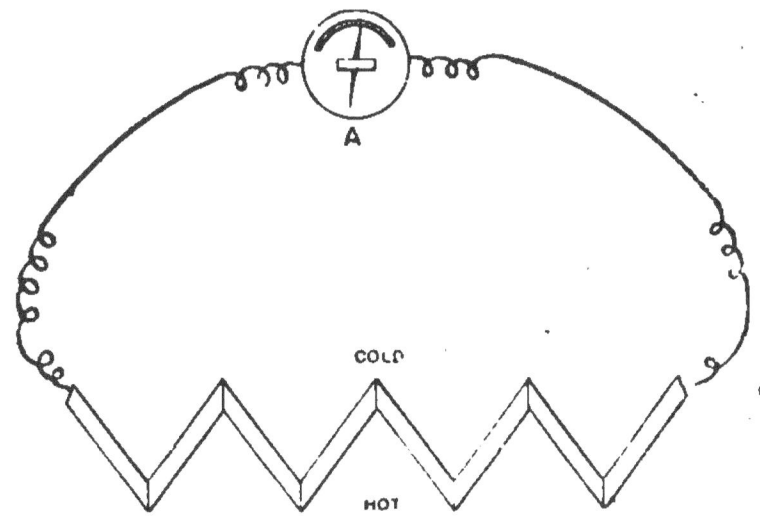

Fig. 90.—Diagram showing Principle of Thermopile.

set is kept cool, an electric stress will be set up in the metals, and this can be transmitted through a wire connecting the two ends of the zigzag as from an ordinary chemical battery. If, therefore, such a zigzag arrangement is made, and a delicate galvanometer A is connected in circuit between the two ends, the needle of this instrument will be deflected, by an electric current, when the set of junctions is heated and the opposite set kept cool.

This necessity of having one set hot and the other cold renders the working of a thermopile a matter of some difficulty, since heated bars of metal soon conduct the heat from one end to the other.

The E.M.F. of each pair in a thermopile is exceedingly low, and although this rises with the temperature of the heated junctions, some idea of its adaptability to practical purposes will be deduced from the

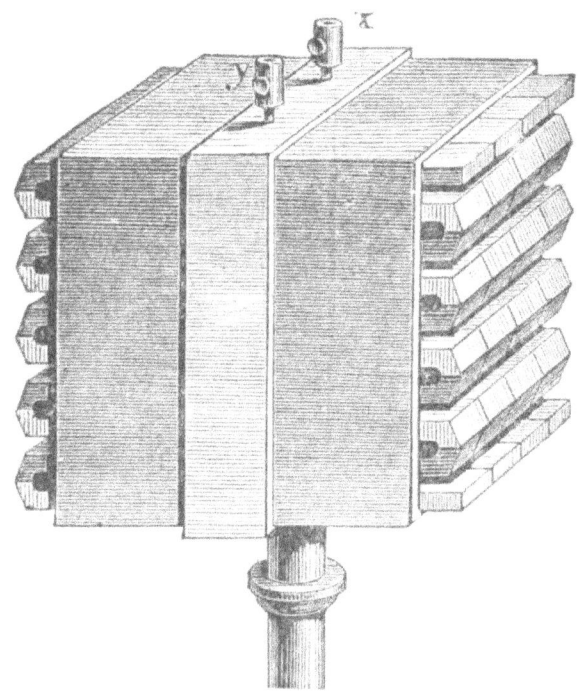

Fig. 91.—A Practicable Form of the Thermopile.

fact that it will take 3,500 bars of iron and copper, with their junctions heated to 212° F. and the opposite junctions kept quite cold to generate a current pressure of 5 volts. These metals are the least expensive, but not the most effective for the purpose. The most effective metals of a dissimilar character are antimony and bismuth ; but these, besides being more costly, are not easily jointed, and the joints offer a great resistance to the current, thus wasting the power generated in the thermopile.

Fig. 91 shows one of the many forms of thermo-piles proposed. Its advantage is that it facilitates a compact arrangement of numerous junctions, as, to increase the E.M.F., either the temperature difference or the number of junctions must be increased, and the latter is obviously more easily effected than the former. It is necessary to arrange that all the odd junctions (Nos. 1, 3, 5, etc.) are on one side and all the even junctions (Nos. 2, 4, 6, etc.) on the other, inasmuch as the junctions have to be heated and cooled alternately. In the arrangement illustrated this is well effected. Insulating pieces are interposed where the metals are not to be in contact. The two ends of the series are joined to the terminals Y X, to which the wires of the external circuit can be connected.

These generators of electricity are of interest to the student in pursuit of knowledge, but are scarcely of commercial value.

INDEX ·

PRINTED BY CASSELL & COMPANY, LIMITED, LA BELLE SAUVAGE, E.C.